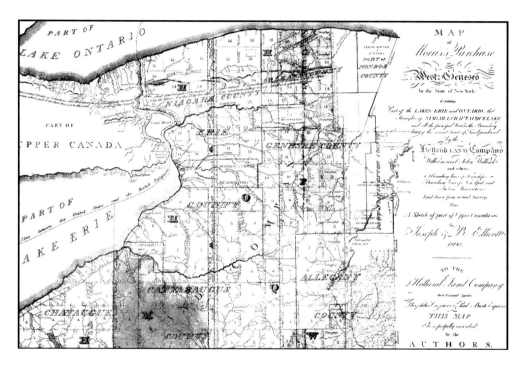

This early map of Western New York was derived from Joseph Ellicott's survey of the "West Genesee" region. It identifies a few early settlements, highways and county lines. Chautauqua, Cattaraugus and Allegany Counties compose the southern tier. *Courtesy of Library Archives, SUNY Fredonia.*

ENERGY & LIGHT

IN NINETEENTH-CENTURY

WESTERN NEW YORK

NATURAL GAS, PETROLEUM & ELECTRICITY

DOUGLAS WAYNE HOUCK

THE
History
PRESS

Published by The History Press
Charleston, SC 29403
www.historypress.net

First published 2014

Manufactured in the United States

ISBN 978.1.62619.300.0

Library of Congress CIP data applied for.

This book is dedicated to James O'Brien, president, and the other officers of the Chautauqua County Historical Society. They are to be commended for their diligence, hard work and success at preserving the history, legacy and memorabilia of Chautauqua County, New York.

CONTENTS

ACKNOWLEDGEMENTS

Preserving our past provides a pathway to the future. We profit from the endeavors, activities, problems encountered and solutions devised by those who have gone before us. Many of the tales, exploits and endeavors included in this book were derived from records, reports, accounts and articles kept on file at the following libraries, archives, and historical societies:

Barker Library—Fredonia, New York.
Buffalo and Erie County Historical Society—Buffalo, New York.
Chautauqua County Historical Society—Westfield, New York
Chautauqua Institution Archives—Chautauqua, New York.
Friends of Drake Well—Titusville, Pennsylvania.
Olean Historical and Preservation Society—Olean, New York.
Pioneer Oil Museum—Bolivar, New York.
Reed Library Archives, State University of New York–Fredonia.

Special appreciation is extended to Niles Dening, board member and photographer at the Chautauqua County Historical Society, for preparing special photos used in this book.

PART I

THE PEOPLE, PLACE AND PROGRESS

WITH MEMORIES AND REFLECTION

I grew up on a Chautauqua County farm in Western New York. One of my first memories is sitting on the floor in darkness, looking up at my father. He was sitting in a chair, bathed in yellow light streaming from a kerosene lamp hanging over his head. This was before rural electrification came along, planting poles and running wire up our road. They had electric lights in nearby Dunkirk and Fredonia, New York, but we didn't have it.

Every morning, Mom collected kerosene lamps. She cleaned chimneys and trimmed wicks. I remember being in the barn in the evening, watching my father, grandfather and Harry, our hired man, milking cows by the light of kerosene lanterns. Darkness was broken by the glow of kerosene.

Some had gaslights. There were gaslights in the Grange Hall. My grandparents had gas, and it came from a natural gas well on the other side of Walnut Creek. I didn't think much about such things in those days, but natural gas, kerosene and, finally, electricity came along. They changed our world, everything in it and how we lived our lives.

History, lore and legends tell us that people have known about petroleum and natural gas for a long time. Noah sealed his ark within and without with pitch, bitumen. Ancient Sumerians used crude oil for smelting, refining and alloying. But the use of crude oil dropped off, declining with passing years. Some believe that the Oracle at Delphi might have been influenced by natural gas coming from faults under the temple. Plutarch reported ancient "eternal fires" in Iraq about two thousand years ago. Even so, there doesn't seem to be any indication that anyone tried to use petroleum or natural gas to generate energy or produce light up until the nineteenth century.

This occurred in Western New York. It happened when William Austin Hart noticed bubbles of gas rising to the surface of Canadaway Creek. He collected natural gas and used it to produce heat and provide light. Sometime later, Dr. Francis Beattie Brewer, a longtime resident of Westfield, New York, initiated the inquiry that resulted in the world's first oil well being drilled near Titusville, Pennsylvania. Charles Brush and Thomas Edison lit electric lights at the Chautauqua Institution, and Nikola Tesla produced electricity at Niagara Falls and transmitted it to Buffalo, New York.

These events—the initial use of natural gas, the initial drilling and use of petroleum and the initial development and use of electricity—occurred one after another, in succession, and in relatively short order in Western New York and nearby northwestern Pennsylvania. This is a small region, somewhat isolated at the time. There were a few small settlements, but there were no cities or urban centers until the later years of the nineteenth century. It was, for the most part, a dense wilderness.

One wonders: How did this happen? Why did it happen here?

A UNIQUE PEOPLE IN AN UNUSUAL PLACE

Using natural gas, petroleum and, finally, electricity to generate energy and create light happened when a unique group of people moved to an unusual place. The people were from the Connecticut River Valley, the nation's first industrial corridor. They brought their knowledge, expectations and experience with them when they moved to Western New York. These people knew how to survive in the wilderness, they knew how to derive products from the environment and they knew how to sell the products they created.

They came from a commercial/industrial background. Being machinists, millwrights and gunsmiths, they were familiar with the sources of energy being used in the Connecticut River Valley—namely, fire and water power. Water power required falling water, captured by waterwheels. Fire, their primary source of energy, was used for smelting, forging, heat and light. Light was created by burning wood, coal and oil—principally whale oil at the time. These people were nineteenth-century entrepreneurs. Creating products, they developed markets. They found them in Western New York.

It was new, but it had a history. Most of the territory beyond the Appalachian and Allegheny Mountains had been claimed by the French for about two hundred years. Great Britain secured it by winning the French and Indian War. England held the land for about twenty years before it became a part of the United States after the Patriots persevered during the American Revolution. Even so, it was disputed territory.

It was claimed by Massachusetts, New York, Connecticut and Pennsylvania. Virginia complicated matters by claiming everything west of the Allegheny River. It took several conferences and an occasional intervention by the federal Congress to work out the state boundaries. However, the New England residents didn't wait. They started moving to the western region in the 1780s. They found it to be an unusual place. It was the world's largest temperate rain forest: watered by heavy, almost continuous, rainfall coming off the Great Lakes. Dense forests covered the hills and valleys.

There were several waterways. Some led to the Great Lakes, and others flowed into the Allegheny River, on to the Ohio and then the Mississippi. Flammable gas occasionally bubbled to the surface of streams. A black, oily substance sometimes coated the surface of springs. It floated away, fouling creeks and rivers.

These newcomers were, for the most part, fourth- and fifth-generation descendants of Puritans, the thirty thousand or so religious dissenters who fled England during the Great Migration between 1620 and 1640. The original Puritans stumbled ashore, in awe of the forest. Most feared the dark, dense woods, as depicted by Hawthorne in his short story "Young Goodman Brown." William Bradford, the governor of Plymouth Colony, described the forest as "a hideous and desolate wilderness, full of wild beasts and wild men."

This changed. The descendants of the Puritans grew up in the forest. They used it. They chopped down trees, built sawmills, cut lumber and cleared the land. Men and women married early. They had huge families, eight or nine children, and continued moving, always heading west. Each generation needed land—more and more land, generation after generation—until they came to, reached and filled the Connecticut River Valley.

The Connecticut River became their thoroughfare; it extended from Long Island Sound through Connecticut to Massachusetts, Vermont and New Hampshire. Early surveyors found iron ore in Connecticut in 1728. John Paul and Ezekiel Ashley discovered "the largest iron deposit in the world" at Salisbury, Connecticut, in 1731. Joseph Skinner started making iron in 1740.

Blessed with the required resources of water power, iron ore, limestone for flux and lumber for charcoal, the Connecticut River Valley Iron industry flourished for several decades before depleting available ore and timber resources.

Plymouth, Vermont, grew up around an early iron foundry and a mining operation for gold, talc, soapstone, marble, asbestos and granite. Prospectors tapped Gardner Mountain's mineral belt for gold and silver. Copper was discovered, and Vermont became the nation's primary copper producer. The Connecticut River Valley, in turn, became the nation's first industrial corridor. Millwrights constructed mills, and craftsmen made potash, iron, steel, copper, firearms and machine tools.

Technology flourished. Tabitha Babbit invented the circular saw; Samuel Hopkins secured the first United States patent, for "the making of Pot Ash and Pearl Ash by a new Apparatus and Process." Eli Whitney invented his cotton gin and started making firearms. Samuel Colt manufactured the revolver, the world's first product with truly interchangeable parts. Eli Terry and Seth Thomas made clocks. Simeon North designed the milling machine. Edward Bullard built the world's first vertical boring mill.

The people were prolific. The population of Connecticut and Massachusetts grew rapidly, more than 7 percent per year for thirty years, from 1760 to 1790. Statistics from the time indicate that most of the rural families living in the Connecticut Valley region raised between seven and ten children, on average, for generation after generation. One county, Hampshire County, the region around Amherst, Massachusetts, can be used to cite the impact of this growth. The population had increased from 6,250 residents in 1765 to more than 76,000 residents by 1810.

Original 160-acre and 200-acre farms were broken up, divided and then subdivided, being cut up into 40-acre plots for children and grandchildren. When the available acreage no longer sufficed, descendants were forced to leave the valley. They sought refuge, seeking opportunity on new lands opening in the west.

The population of Hampshire County, Massachusetts, dropped off dramatically. It plummeted from 76,275 residents in 1810 to 26,487 residents in 1820, a sudden 65 percent drop in population. For the most part, those leaving the valley went west. Jumping over the Hudson Valley Dutch, they trekked through Central New York, seeking new homes in Western New York and northwestern Pennsylvania.

It was a different world—the great forest extended for hundreds of deep, dark miles, and the ground was littered with fallen forest giants. There were bogs, briars, lakes and waterways. The forest was too dense, too thick, for riding a horse. Travelers had to walk, picking their way between trunks of gigantic trees and over and around fallen logs. Branches cluttered the ground. Winters were severe, marked by swirling snow. There weren't many roads. British gunboats patrolled Lake Erie. Wilkinson ran a salt boat on the lake, but that was about it.

American commercial vessels didn't much venture on Lake Erie and the other Great Lakes until after the War of 1812. The *Walk-in-the-Water* was the first steamboat on Lake Erie. It was built by a group of enterprising New York businessmen. The group included Noah Brown, who designed the craft, and Robert McQueen, a machinist who designed and built the engine. They started building their steamboat at Black Rock on the Niagara River in 1816, but they didn't get it launched until 1818.

The *Walk-in-the-Water* was considered a marvel. It got a good seven miles to a cord of soft wood, most of the time. It could, when running with the wind, sometimes reach the astounding speed of five or six miles in a single hour. Flatboats and keelboats were used on the Allegheny River.

Letters could be sent by the postal service, but they took weeks to get anywhere. Richard Williams, an early settler, secured a contract to carry mail to and from Buffalo. When sick or otherwise indisposed, his wife, Sophia, made the run. Legends persist of Sophia swimming her horse across the Cattaraugus, Eighteen Mile and Buffalo Creeks holding the mail above her head and delivering it on time.

People had to be self-sufficient. They built their own dwellings. They obtained and raised their own food. They used whale oil for lighting back in New England, but whale oil was scarce in Western New York. It was expensive, so most couldn't afford it.

The few "well-to-do" settlers in Western New York used candles, but most couldn't afford such a luxury. They had to rely on the flickering illumination obtained from their fireplaces. That was the situation until William Hart, a transplanted Connecticut gunsmith, came up with an alternative. He initiated a progression of energy and light.

CHANGE: A PROGRESSION OF ENERGY AND LIGHT

"Change," grandfather insisted, "happens slowly. It sneaks up on you when you're not looking." Clayton was born in 1871 and lived in Chautauqua County most of his eighty-five years of life. He told me, on several occasions, that he had seen, experienced and lived through a lot of change in his life but that it came along from time to time, one event after another. He said it started with the bicycle. A few brave fellows started pedaling around on their high front-wheel bikes, "penny-farthings" and "ordinarys," when he was a kid. This gave rise to headers and highways.

Women joined them, riding their new safety bikes. Ladies had to give up their voluminous skirts. Grandfather said the world was better for it. The telephone arrived. You held a thing to your ear, stood in front of a box and yelled at it—after an operator connected you. Automobiles chugged along in due course, followed by the airplane. Radio arrived and then television.

That was the beginning. They made pursuit planes and cargo aircraft at the Curtiss-Wright plant in Buffalo when I was growing up. They made Moon Lander Research Vehicles at Bell Aero-Systems in Tonawanda a few years later. Computers once filled entire rooms. More sophisticated computers are now held your hand. Our sky is filled with aircraft. Hordes of automobiles crowd our streets and highways. Expressways span continents. New products appear by the hour. Distance is measured by hours rather than by days or miles. Gasoline is a concern, and fracking is an issue.

Change may have started slowly, but the pace of change has accelerated, driven by a proliferating technology. However, when viewed in the grand total perspective of more than 200,000 years of human existence, this thing called "change" appeared abruptly; it occurred suddenly, without warning, in a relatively short period of time. Our current mode of living, our contemporary lifestyle, constitutes a complete, absolute break from the multi-thousand-year norm of human existence. Moses had to walk from Sinai to Egypt to confront pharaoh, and George Washington walked much of the way to Fort Le Boeuf to confront the French.

Then change started. Thomas Slavery is given credit for developing the world's first steam engine. This happened yesterday—at least yesterday in the grand span of our human existence. Slavery patented the engine in 1698, a little more than four hundred years ago. Slavery used his machine to pump water from mines. It worked by injecting steam, under pressure, into a closed vessel filled with water; this forced the water upward, out of the shaft. A cold sprinkler then condensed the steam, creating a vacuum. This sucked more water from the mineshaft.

Thomas Newcomen improved Slavery's engine in 1712. His "fire engine" worked by injecting steam into a cylinder fitted with a piston. The steam pushed the piston, and the piston pushed a rocking arm downward. Cold water was then injected into the cylinder, cooling the cylinder. This created a partial vacuum, pulling the piston and arm back to their original position. The fire engine generated twelve strokes per minute, every minute lifting ten gallons of water with each stroke. Strength was estimated to be about 5.5 horsepower—not impressive by today's standards, but it was considered a sensation at the time.

James Watt improved the engine in 1769, coming up with a separate condenser connected to the cylinder. This avoided cooling the cylinder and was more efficient. Watt's engine soon became the dominant design. This occurred about 250 years ago.

Robert Fulton put a steam engine on a boat, the *Clermont*, about two hundred years ago. The world's first successful steam locomotive, the *Rocket*, lurched down the first track in Great Britain about 180 years ago. Each event, each discovery, facilitated another. The pace of change accelerated with the discovery and use of fossil fuels. Coal was the first fossil fuel.

Ancient records tell us that the Chinese have used coal for heating and cooking for about four thousand years. Marco Polo mentioned this in his *Description of the World*. The English started using coal to power steam engines in the 1700s. The *1761 Plan of Fort Pitt and Parts Adjacent* reports coal mining taking place in North America. Other curiosities were noted.

Early French explorers came upon an oil spring in what is now Allegany County in Western New York. They noted it, but nothing was done with it. Settlement of the region did not begin until after the French and Indian War, and then the American Revolution, had ended. When settlers started entering the region, they came from the Connecticut River Valley.

The settlers spilled over the hills. They crossed the Hudson River and headed west. Many of them headed for the Allegheny Basin, the 11,580-square-mile region drained by the Allegheny River. There were conflicting land claims, but that didn't seem to bother the settlers. They came anyway and kept coming for about thirty years.

The colonial charters of Massachusetts and Connecticut extended the boundaries of the colony all the way across the continent to the South Sea or Pacific Ocean. New Yorkers believed that their boundary extended to the western end of Lake Ontario. Moreover, Virginia claimed everything west of the Allegheny River. Pennsylvania's western boundary was uncertain and undefined. Penn's original charter extended his colony's western boundary from some uncertain point in dispute.

Two states, New York and Massachusetts, claimed the land between Seneca Lake and Lake Erie, now Western New York. Delegates from New York State and Massachusetts met in Hartford, Connecticut, in 1786 to resolve the issue. After considerable negotiation, they agreed to divide the rights. Massachusetts got preemption rights: the right to sell the land. Once the 8,973-square-mile region was sold, it would become part of New York State.

Massachusetts sold the land to Oliver Phillips and Nathaniel Gorham for $1 million, payable in specie or Massachusetts securities. Massachusetts used the money to pay off some of the state's war debt. When Phillips and Gorham fell behind on their payments, they sold about half their purchase, 3.3 million acres, to Robert Morris, a financier of the American Revolution. Robert Morris, in turn, sold the land to a group of Holland merchants. This Holland Land Company then bought another 1.4 million acres of land in what is now western Pennsylvania, on and around the Allegheny Basin, including what is now Warren, McKean, Forest, Crawford and Venango Counties.

New York State law complicated matters by prohibiting a foreign enterprise, the Holland Land Company, from owning land in the state. The company then hired Aaron Burr, a New York City attorney, to represent it and try to change the law. As Burr happened to be a member of both the New York State legislature and United States Senate, he was able to do this. Even so, the Holland Land Company had to hire a general agent to conduct business in the United States.

The land company hired Theophilus Cazenove, a merchant and banker from Amsterdam who happened to be living in Philadelphia, to be its agent. Needing to know what was included in the purchase, Cazenove hired Joseph Ellicott to run a survey. Ellicott spent two years (1798–1800) running his survey. He lived outside in summer and winter, running transit lines and laying out townships. After finishing the survey, Ellicott prepared a detailed map or chart of the purchase.

Cazenove then picked Ellicott to be his Western New York agent. Ellicott set up a land office in Batavia, New York. He served as the Holland Land Company's Western New York agent until 1821. Supervising land sales, Ellicott laid out Batavia, New York, and the village of Buffalo. He selected mill sites and cut roads and helped plan the Erie Canal. As seller and agent, Ellicott offered generous terms to buyers; some purchased land for as little as twenty-five cents down.

Connecticut Valley residents rushed to take advantage of this opportunity. They jumped over the Hudson Valley Dutch and moved through Central New York to purchase property in Western New York. Many of the Hudson Valley Dutch looked on this intrusion with dismay, expressing concern and misgivings. The Dutch came up with a name for the interlopers. They called them "Jan Kees."

The Dutch *j* being pronounced as the English *y* resulted in the name becoming uttered as "Yankee." The name stuck and has since been applied, first to people from the northern states and, more recently, to people from the United States in general. There are various interpretations or explanations for the name—the most acceptable perhaps being the literal Dutch, "little John."

New England residents had their own feelings. Estwick Evans, a New Hampshire traveler, noted in his 1819 travelogue, "The variety of people in Albany is great. The Dutch here still make a considerable figure; but Americans are becoming more numerous." By "Americans," Evans meant those of English heritage, primarily from the New England.

Most of the newcomers took the Great Genesee Road. This early highway, graveled for the most part, ran from Utica, New York, to Geneseo, a settlement on the Genesee River. There was traffic, lots of traffic every spring. During the early years, many turned south after they got to Geneseo. Olean Point, a settlement on the Allegheny River, was their destination. The Allegheny River became a major waterway; it was an open route, a highway to the Ohio River and beyond. Congress declared it a public highway in 1807.

Westbound traffic dropped off during the War of 1812. The British burned Buffalo and the entire Western New York region along the Niagara River. The Holland Land Company responded by opening (cutting out) a new Chautauqua Road from the head of Chautauqua Lake to connect with the old Olean Road at Ischua, in the Ischua Creek Valley. This was about twenty miles north of Olean Point.

Travelers now had a choice: they could take the old road to Olean and the Allegheny River, or they could take the new Chautauqua Road. This road took them to Chautauqua County, Lake Erie and New Connecticut, Connecticut's Western Reserve. Even so, many continued going to Olean and the Allegheny River.

The trickle of westbound traffic during the War of 1812 turned into an absolute flood after the war ended and when 1816, the year without a summer, occurred. Andrew Young provided a month-by-month description of the dreadful weather that occurred during that year:

> *March 1816 was cold and boisterous, the first half of it; the remainder was mild.*
>
> *April began warm, and grew colder as the month advanced, and ended with snow and ice, with a temperature more like winter than spring.*

May was more remarkable for frowns than smiles. Buds and fruits were frozen; ice formed a half inch in thickness, corn was killed; and the fields were again and again replanted, until deemed too late.

June was the coldest ever known. Frost and ice and snow were common. Almost every green thing was killed; fruit nearly all destroyed. Snow fell to the depth of ten inches in Vermont and Maine; three inches in New York. It also fell in Massachusetts.

July was accompanied by frost and ice. On the morning of the fourth, ice formed of the thickness of a common window glass, throughout New England, New York, and parts of Pennsylvania. Indian corn was nearly all killed; few favorably situated fields escaped. This was true of some of the hills of Massachusetts.

August was more cheerless, if possible, than the summer months already passed. Ice was formed a half inch in thickness. Indian corn was so frozen that the greater part was cut down and used as fodder. Almost every green thing was destroyed in this country and Europe. Papers from England said, "It will ever be remembered by the present generation, that the year 1816 was a year in which there was no summer." Very little corn in New England and the middle states ripened; farmers supplied themselves from corn produced in 1815 for seed in 1817. It sold for $4 to $5 a bushel.

September furnished about two weeks of the mildest weather of the season. Soon after the middle, it became very cold and frosty; ice forming a quarter of an inch in thickness.

October produced more than its usual share of cold weather; frost and ice common. November was cold and blustery; snow fell to make sleighing. December was mild and comfortable. Very little vegetation matured in the Eastern and Middle states. The sun's rays seemed to be destitute of heat throughout the summer; all nature was clad in sable hue; and men exhibited no little anxiety concerning the future of this life.

We now believe that this unusual weather event was caused by a volcano, Tambora, blowing up on the other side of the world. Nobody knew what caused the terrible weather at the time. Hundreds and then thousands of New England and northern New York residents fled their homes. Rushing through Central New York, they sought the warmer climate and longer growing season farther south on the Allegheny River. These were a unique people, and they were moving to an unusual place.

PART II

DERIVING PRODUCTS
FROM THE ENVIRONMENT

Lamps, Light and Lighting

Early man lived by the light of the sun and endured hours of darkness. Polynesian mythology tells us that Maui journeyed to the top of Haleakala to snare the sun and delay it. Greek mythology claims that Prometheus suffered the wrath of the gods to bring mankind the gift of fire. Fire became the illuminant. Archaeologists now tell us that human beings have been using and controlling fire for at least 125,000 years.

The first oil lamp appears to have been invented about seventy thousand years ago. A hollow rock, shell or other natural found object was probably filled with moss or similar material soaked with animal fat and ignited. Humans then started imitating these natural shapes with pottery, alabaster and metal, creating lamps. Wicks were later added to control the rate of burning. About nine thousand years ago, Greeks started making terra-cotta lamps to replace their hand-held torches. The word *lamp* comes from the Greek word *lampas*, meaning torch.

The Egyptians developed a crude form of candle about seven thousand years ago. They used rush lights, or torches, made by soaking the pithy core of reeds in molten tallow. Their candles didn't have a wick. Romans are credited with developing the first wick candles, using them to aid travelers,

Left: Properly furnished parlors in nineteenth-century homes placed the kerosene lamp on a marble-top table. A stereoscope on a chair was an added feature. The lamp provided light; the stereoscope provided entertainment. *Courtesy of Yorker Museum. Photograph by Niles Dening.*

Below: The old "Yellow Dog" was a double-wicked lamp used to provide illumination. It was first used on the yardarms of whaling ships, being filled with whale oil and set alight. These lamps were later used in drilling rigs in Western New York and northwestern Pennsylvania to illuminate evening drilling operations.

light homes and illuminate places of worship during hours of darkness. Like the Egyptians, the Romans relied on tallow, gathered from cattle or sheep suet, as the principal ingredient of their candle.

Europeans started using beeswax, a substance secreted by honeybees making honeycomb, to make candles during the Middle Ages. Beeswax candles were a marked improvement over the older tallow candles. They did not produce a smoky flame or emit an acrid odor. Beeswax candles burned pure and clean, but they were expensive. Only the wealthy could afford to use them.

Colonial women made America's first contribution to candle making. They discovered that boiling the grayish-green berries of the bayberry bush produced a sweet-smelling wax that burned clean. The first written account of bayberry candles appeared in 1698. It proclaimed that instead of "stinking…they perfume like incense." Extracting wax from bayberries was a time-consuming task, but bayberry candles were in great demand until the Argand lamp became available.

Ami Argand, a Swiss chemist, is credited with developing an oil lamp with a circular wick surrounded by a glass chimney. This occurred about 230 years ago, in 1783. His basic idea was to have a cylindrical wick that air could flow through and around. This increased the intensity of the light. A cylindrical chimney enhanced the airflow, and a series of experiments provided the proportions for optimum operation. A mechanism for raising and lowering the wick allowed some adjustment and optimization as well. The light was much brighter than a candle, by a factor of five to ten. The lamp burned cleanly, and it was cheaper than using candles.

Argand had to solve a number of problems while developing his lamp. A lace maker came up with the design for the wick. Glassmakers created a clear glass for the chimney strong enough to withstand the heat of the flame. All of the available types of oil were tested to find one that produced a clear, odorless flame. Whale oil was eventually selected. This created two new industries: lamp making and whaling.

Argand-style whale oil lamps were originally manufactured in England and France. Famous Americans such as Benjamin Franklin, Thomas Jefferson, John Adams and a few others who traveled to Europe took note of the Argand-style lamps and their bright light. Being impressed, they brought Argand lamps home with them. Philadelphia craftsmen started making them. John Leadbeater, Peter Geley and Christian Cornelius were the most prolific. They produced hundreds of Argand-style whale oil lamps. Then local jewelers, tinsmiths and copper plate makers took notice and started making their own lamps. Argand-style lamps, burning whale oil, became the lamp of choice for those who could afford them.

Whaling ships started sailing out of Nantucket and New Bedford, Massachusetts, to get the oil. Some have suggested that more than two hundred ships might have been out at any given time in 1834. There was a reason: whaling produced profits that were unthinkable at the time. A single ship, out for thirty months, could generate a $35,000 profit. Americans dominated the whaling industry. Ships sailing out of New Bedford and Nantucket during the 1840s and '50s cleared more than $10 million per year, each year, for about twenty years. That's in 1840 dollars. The country's entire domestic output was about $3 billion per year.

Whaling became an early indicator of an emerging New England business ethic: conduct an enterprise, pursue advantage and use it to create wealth. During its peak year, 257 ships came home, bringing back 431,000 barrels of oil. The average price for a gallon of whale oil rose from $0.32 to $0.88 over the twenty-year period between 1830 and 1850. For the more expensive sperm whale oil, the price increased from $0.65 per gallon to $1.80 per gallon. A skilled tradesman at that time might have been able to generate between $100.00 and $150.00 per year—providing the tradesman was able to find enough work to keep him busy for the entire year.

Whale oil was expensive. Most did without.

PRODUCTS FROM THE ENVIRONMENT

New England craftsmen became adept at creating products, deriving them from their environment. Their rugged environment didn't offer much in the way of resources—it had rocks, trees and cold weather. Most merchants wanted gold, molasses, rum and sugar. Finding a way to trade the limited resource at their disposal for the items they desired required considerable creativity and some ingenuity. They had that and ice to spare. Frederic Tudor was the one who came up with a unique idea: trade the ice. He made a fortune doing so, becoming one of America's first millionaires.

Frederic's grandfather, Deacon John Tudor, came to Boston in 1715. Frederic's father, William Tudor, graduated from Harvard in 1769 and served as a judge advocate in the Continental army during the American Revolution. He returned home to Boston at the end of the war, set up a law practice, married Delia Jarvis and became a judge. His son, Frederic, was born on September 4, 1783.

The Tudor family assumed that young Frederic would probably follow in his father's footsteps, but Frederic rejected college. He opted, instead, for a business career. When he was thirteen years old, he went to work as an apprentice at Ducator & Marshall, a State Street store in Boston. Four years later, Frederic took a trip to Havana, Cuba, with his brother, John. Engaging in a few business dealings there, Frederic discovered that Cuba had a lot to offer, including mahogany furniture, coffee, molasses and sugar.

Sailing back to Boston, Frederic came up with a plan. He'd sell ice, shipping it to the tropics. Ice would be a rarity. He believed that it could be sold or traded for coffee and sugar. Ice could be cut from Massachusetts ponds, stored in icehouses, put on board sailing vessels insulated with saw dust and shipped to the Caribbean. Frederic launched his first vessel on February 13, 1806, sailing to Martinique. Ice was indeed a novelty; nobody knew what it was or what to do with it, but they bought it.

Frederic sent a second cargo of ice to Havana in March 1806. A third followed that April. All of his shipments arrived safely, with the ice in good condition. Loading the ships with molasses for the return trip, they sailed back to Boston. Frederic sold the molasses in Boston for a substantial profit. Frederic followed up by going back to Havana in 1807 to build a better icehouse to store his ice after it arrived. This worked and business improved. Then Frederic thought up another marvelous idea. He started advertising his product as "ice with quinine and rum: sure fire tonic to prevent malaria." People started buying ice, putting it in drinks. We still do.

Trees were another resource. Some could be sawed to make lumber. Hardwood could be cut for firewood and the ashes gathered to make potash. There had been a growing industrial demand for potash for some time. The Industrial Revolution in England demanded ever-increasing quantities of potash for the production of textiles, soap, dyes, baking powder, gunpowder and other products. Potash was derived solely from the ashes of hardwood trees until 1860.

As England and Western Europe had cut their hardwood forests long ago, all of the potash and hardwood ashes required for European industry had to come from North America. England needed several thousand tons of potash every year. British industrialists responded to this need in 1760 by investing large sums of money building elaborate potash works in Boston, New York City and Philadelphia. The works were equipped with elaborate leaching vats, furnaces for evaporation and other equipment.

The Boston works produced large quantities of potash from hardwood ashes brought in by New England residents and farmers. Ashes gathered along the Mohawk and Hudson River Valleys were shipped down the Hudson River to New York City. The Philadelphia works was not as productive as southern Pennsylvania lacked the hardwood forests needed for potash production.

Manuals on potash production were distributed to the northeastern settlers, and experts were sent into the rural areas to offer advice and assistance on potash production. Black salts, a concentrated form of crude potash, were produced on farms in isolated rural areas, packed in wooden barrels and transported to the nearest market. It was easier to ship barrels of black salts than bushels of dry ashes.

Samuel Hopkins bought land near Pittsford, Vermont, in 1781 and went into the potash business. Black salt was made by burning hardwood logs, such as elm, ash, maple, hickory, beech and basswood. The ash was collected and placed in wooden vats. Water was poured over the ash and the resulting gray-brown liquid cooked in big iron kettles over open fires. The moisture was boiled off. The remaining block of cooked "pot-ash" contained particles of carbon, forming "black salts." The demand for the thick-walled iron kettles became so great that Ethan Allen and his brothers went into the kettle business. They started manufacturing potash kettles at their Salisbury, Connecticut ironworks.

In 1790, Samuel Hopkins received the first United States patent ever issued for his improvement "in the making of Pot Ash and Pearl Ash by a New Apparatus and Process." The patent granted Samuel Hopkins exclusive rights to the use and merchandising of his process and product for fourteen years. The Hopkins process required burning raw ashes in specially constructed brick ovens before the ashes were dissolved in water. The second burning of the ash resulted in a much greater carbonate formation, apparently because the free carbon in raw ashes (which partly explains the black color) was more completely oxidized and because of exposure to the carbon dioxide gas from the fire.

Yields were also increased by mixing the insoluble residue from one batch with the raw ashes of the next. As the process required the construction of asheries subject to the conditions of his patent, Mr. Hopkins sold the rights to use his process and construct ovens by issuing five-year licenses for the use of the process and apparatus. Each license required an initial payment of $50 and another $150 paid out in five annual installments. Enterprising merchants rushed to buy licenses. As much of the timber in the Connecticut

River Valley had already been cut, the industry flourished in the new lands opening up in Central and Western New York.

Major Samuel Forman of Cazenovia, New York, was one of the first merchants to purchase a license for the operation of an ashery using the Hopkins process and apparatus. He had previously worked as a clerk in the mercantile firm of Ledyard and Walker in New York City before moving upstate to open his store and ashery. The store provided the settlers moving into the area with the hardware and supplies required for survival, and the ashery provided the settlers with a little cash that could be used to make the payments on their debt at the store.

The Cazenovia ashery manufactured pearl ash and packed it in wooden barrels. The wooden barrels were taken to the Mohawk River and placed on flatboats for the trip down the Mohawk and Hudson Rivers to Ledyard and Walker in New York City. Ledyard and Walker loaded the barrels of potash into the holds of ships bound for London.

The barrels served as ballast for the trip across the Atlantic, and the potash was sold for $400 or $500 per ton when it got to London. Ledyard and Walker received 2.5 percent of the proceeds for their part in the transaction. A portion of the remainder was applied against the cost of the merchandise being sent up the river to stock the store, and the rest of the money was banked in Mr. Forman's account at the banking house of Prime, Ward and King in New York City.

Jacob Ten Eyck left Albany in 1798 and moved to Cazenovia to take a position as a clerk in Major Forman's store. Jacob worked at the store and ashery until he took over management of the ashery in 1807. Jacob then expanded the business by setting up satellite stores and asheries throughout Madison County and the surrounding region of Central New York State. As local forests were cut away, Jacob expanded his business, moving to Western New York.

Jacob sent a shipment of merchandise to the Plumb brothers' store, a new store opening in far-off Chautauqua County in 1816. The store, the first in Chautauqua County, was located in the small settlement that would later become Fredonia, New York. The settlement was located on flats along Canadaway Creek, about two miles from the Dunkirk Harbor on Lake Erie.

Jacob sent his shipment out by freight wagon in June, over the Genesee Road to Black Rock, a community on the Niagara River. Due to inclement weather, the shipment didn't get to Black Rock until the end of August. The merchandise was then loaded on the lake schooner *Kingbird* for a short voyage to Dunkirk Harbor. The schooner sailed into Dunkirk Harbor in the middle of September. The merchandise was unloaded and hauled to the store.

However, the Plumb brothers ran into a few problems. They traded their merchandise for ashes but failed to process the ashes. Ashes were brought in, but they were left outside, uncovered. Rain and snow washed away the potash. Growing impatient, Jacob sent his nineteen-year-old clerk, Walter Smith, on a mission to collect the debt. Walter was originally from Wethersfield, Connecticut. Leaving home when he was about fifteen, he went west to seek his fortune. Jacob hired him, took him in and gave him a job.

After the Plumb brothers defaulted on the debt, Walter took over their store and ashery in Fredonia. He ended up clearing $25,000 during his first year of operation. However, the "wood choppers" chopping down the trees, burning them, collecting the ashes and making black salt didn't do as well.

William Peacock, Chautauqua County's subagent of the Holland Land Company, commented on the enterprise, "One acre of hard timbered land dominated by maple trees can yield up to five hundred pounds of black salts." Beech trees weren't that good; they could only yield about three hundred pounds. Wood choppers and ash burners could usually get a few cents per bushel for ashes and from $2.50 to $3.00 per hundred pounds for black salts. Clearing one hundred acres of hardwood could generate $400.00 or $500.00.

Settlers were able to take out an article on a section of land, 15 percent down on a section (640 acres) at $1.00 to $1.50 per acre. The rest could be paid off over a ten-year period. Thus, an enterprising "wood chopper" could purchase a section for about $100.00, clear the land, saw the softwood (pine and hemlock) into lumber and burn the hardwood (maple, beech and ash) for ashes. When that was done, the wood chopper could try to sell the cleared land to somebody else or could just walk away, abandon the property and move on. In 1822, Paul Busti, general agent and director of the land company, complained, "The purchase is overspread with woodchoppers."

Walter Smith, however, prospered. He moved his business from Fredonia, New York, to the Dunkirk Harbor; purchased several lake schooners; and expanded his potash business, moving it to Michigan and then to Illinois as "wood choppers" cut the Western New York forest. Walter developed two lake ports—one became Chicago, Illinois, and the other became Toledo, Ohio.

Sailing the Great Lakes was risky business. There were few beacons, landmarks or other navigational aids on the lakes in the 1820s. In his usual fashion, Walter dealt with this problem by setting aside a parcel of land

Dunkirk's second lighthouse was built in 1878. It stands on land donated by Walter Smith. This lighthouse was lit by kerosene. *Photograph by Hoffman.*

on Point Gratiot to construct a lighthouse. He hired Jesse Peck to be the contractor, and Jesse put up the lighthouse tower in 1826. Walter wanted to light his lighthouse with natural gas, so he hired William A. Hart, the Fredonia mechanic and gunsmith, to run a gas line two and a half miles from Fredonia to the lighthouse.

William Hart constructed a wooden pipeline using pump logs and ran it from Fredonia to Walter's lighthouse, but he encountered a problem. Fredonia is about 150 feet higher than Dunkirk. As natural gas happens to be lighter than air, they weren't able to get it to go down the pipeline. Walter gave up on this first attempt to light a lighthouse with natural gas. Surrendering to forces beyond his control, Walter purchased a whale oil lamp, whale oil and a reflector for the lamp.

Walter's lighthouse, the first Dunkirk Lighthouse, guided lake boats in and out of the Dunkirk Harbor for more than fifty years before being replaced by the second lighthouse in 1878. This lighthouse was lit by kerosene. The second Dunkirk Lighthouse still stands. It is open to the public as a memorial and lighthouse museum.

Later on, Walter Smith became the mastermind and principal proponent for the construction of the world's first long-distance railroad, the New York and Erie Railroad. This early railroad ran through the southern-tier counties of New York State, connecting Piermont on the Hudson River with Dunkirk on Lake Erie. Passing through Olean, New York, the Erie railroad happened to be in the right place at the right time to ship great quantities of oil once the Bradford and Allegany County oil fields came on line.

Olean, New York, produced 80 percent of the world's oil for a few years during the 1880s.

Western New York and Allegheny Basin

Large portions of Chautauqua, Cattaraugus and Allegany Counties in Western New York are in the Allegheny Basin, drained by the Allegheny River. The river begins at a little spring on a hillside in Potter County in north-central Pennsylvania. From there, the river flows northward into New York State before looping back into Pennsylvania.

The Allegheny River is about 325 miles long; it flows generally southward to join the Monongahela River at Pittsburgh, creating the Ohio River. It drains an 11,747-square-mile area of southwestern New York State and northwestern Pennsylvania. The New York portion includes six lakes. Natural lakes, remnants of the last ice age, include East and West Mud Lakes, Bear Lake, Cassadaga Lake and Chautauqua Lake. Chautauqua Lake is largest, about twenty-two miles long.

Cuba Lake, in Allegany County, is an artificial lake that was created as part of a water storage and feeder system supplying the old Genesee Valley Canal. The Genesee Canal, built in the 1840s, was dug to connect the Erie Canal with the Allegheny River at Portville. The Seneca Oil Spring is located at the southwest tip of the Cuba Lake.

The Allegheny River drains most of the northwestern portion of Pennsylvania. This was disputed territory. Connecticut claimed most of the region before the American Revolution, based on its charter issued by King Charles II in 1662. However, Pennsylvania had another charter. King Charles II gave it the same land in 1681. Perhaps the king got confused; however, maps weren't that good at the time.

Pennsylvania got the land, by court decision, after the Revolution. The commonwealth divided this territory into two large counties: Northumberland County and Allegheny County. Settlers started arriving after the Revolution ended, many coming from the east, the Connecticut River Valley. The Allegheny River beckoned: it provided a broad waterway to the south and west, into the heart of the continent.

However, the newcomers discovered that they were in a very different world. Trees were bigger, and there were more of them. Ancient giants of the temperate rain forest had been growing unmolested for hundreds, even thousands, of years. The huge forest was watered, almost continuously, by prevailing moisture-laden winds blowing over and coming off Lake Erie.

Samuel Dale's 1815 survey of the Allegheny Basin provides the following picture of the rain forest: 30 percent beech, 27 percent hemlock, 8 percent sugar maples, 6 percent birch, 6 percent chestnut, 6 percent white pine and 5 percent red maple. This accounted for 88 percent of the growth. A more extensive survey record described a forest that was 43 percent beech, 20 percent hemlock, 6 percent birch, 5 percent sugar maple, 5 percent red maple, 4 percent white oak, 3 percent white pine and 3 percent chestnut. Dry sites were predominantly oak, red maple, white pine and chestnut. Moist areas had hemlock and mixed hardwoods, especially beech. Sugar maples were especially abundant in better-drained ridges possessing sandstone-derived soil.

While growing up, I had the good fortune of walking through and camping in one of the last stands of old-growth forest left in Chautauqua County. The forest was located some distance behind an elderly gentleman's farmhouse on a hillside in the town of Arkwright. Water flowing from springs in his "never chopped" preserve formed a little stream that became one of the sources of the Conewango Creek, a tributary of the Allegheny River. The Conewango joins the Allegheny at Warren, Pennsylvania.

It took five teenagers, arms outstretched, to reach around the trunk of one massive hemlock tree. We couldn't see the top. There were a lot of big trees: hemlocks, maples, beech trees and cucumbers. Unfortunately, the old gentleman passed away, and "wood choppers" rushed to his parcel. They cleared the land. The stream dried up, departing with the trees.

However, a twenty-acre parcel of old-growth forest is left in Chautauqua County. This forest is being maintained and preserved by the Lily Dale

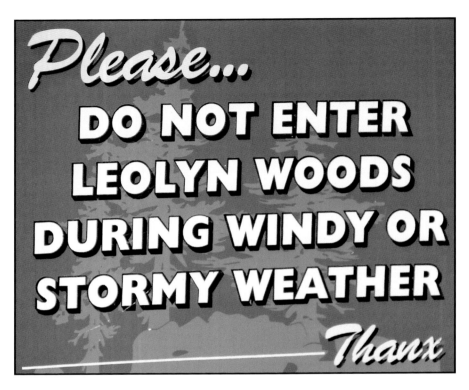

This welcome sign greets visitors entering Leolyn Forest on the assembly grounds at Lily Dale, New York. *Photograph by Hoffman.*

Assembly on its grounds on Cassadaga Lake. The Lily Dale Assembly is located about seven miles south of Fredonia, New York. This forest has a name. It's called Leolyn Woods.

Walking the graveled pathway encircling Leolyn Woods provides a glimpse of the past, but the forest, for the most part, lacks the springs, bogs and marshes that characterized the original area. Several fallen forest giants remain where they fell; however, a few have been cut with chainsaws to maintain the integrity of the walkway. It's an interesting walk, although a few sights and sounds of the present era intrude along the pathway around the edge of the forest.

There's an opening, a meeting place, in Leolyn Woods with benches and an Inspiration Stump. Programs are offered in the forest during Lily Dale's summer season, which usually runs from the end of June until the first of September. The Lily Dale Assembly was established in its current site on Cassadaga Lake by the Spiritualist Association in 1879.

Tall trees tower over and around those who walk the graveled pathway encircling Leolyn Woods. *Photograph by Hoffman.*

Fallen logs and broken branches litter the old-growth forest floor. They are a feature of Leolyn Woods and remind us of the difficulties experienced by early travelers as they made their way through the rain forest of Western New York. Waterways were important; they became highways. *Photograph by Hoffman.*

Cassadaga Lake is actually three small lakes linked together. They empty into Cassadaga Creek, which flows into the Conewango Creek. Early settlers used the Conewango Creek and Allegheny River as thoroughfares for many years. Walter Smith and a few others chopped out a portage road linking Fredonia, New York, with Cassadaga Lake in the early 1800s. Then they tried to clear Cassadaga Creek for flatboats and keelboats. However, it appears that they had limited success.

Many early roads were cut to connect waterways. These first roads were chopped through the forest and under and around the trees. During the summer months, light filtered through the unbroken forest canopy that stretched from the Hudson River to the Great Lakes and beyond. Fall was brilliant with sun and falling leaves. Winter was a time of ice and snow. For several years, the forest prevailed.

Settlers chopped down trees to build cabins and clear fields. Using a "cut and burn" technique, trees were chopped off two or three feet above the ground. Trunks were then cut into logs and dragged away; some were used for construction. Hardwoods were piled to dry and burned. Ashes were gathered and cooked in potash kettles.

Settlers tried to settle on or near waterways. Hemlock and pine trees could be cut and floated downstream to a water-powered sawmill. White pine logs, often 170 feet long, were in great demand. Other logs were split into firewood. Thousands of cords of firewood were floated down the Allegheny River to settlements along the river. Cattle were raised and left to forage in the forest.

Cooking iron became a way to make money. Ore was present, and iron making was relatively simple. Three tons of iron ore were mixed with 190 bushels of charcoal and three hundred pounds of limestone. This was dumped in a huge, vertical stone furnace and ignited to yield a ton of pig iron. Charcoal furnaces sprang up throughout the Allegheny Basin. Metalworkers floated pig iron down the Allegheny River on barges and boats to Pittsburgh. This is how Pittsburgh's iron and steel industry got started.

A black, tarry substance was sometimes noted. It oozed from the ground, gathered on the surface of springs and floated off, fouling creeks and rivers. The Senecas gathered this oil, using it as medicine. Perhaps it had value. Maybe it could be used for something. Flammable gas sometimes bubbled to the surface of streams. It was a peculiar place.

The Allegheny Basin ended up being divided, split between two states, New York and Pennsylvania. Settlers were, for the most part, the same. They shared a common heritage and enjoyed a similar lifestyle. However, there is a difference. Pennsylvania spells the river's name *Allegheny*. New York spells it *Allegany*.

CHAUTAUQUA COUNTY, NEW YORK: LOCATION AND EARLY YEARS

Chautauqua County is New York State's far western county. It's the southwestern corner of the state. Being situated on Lake Erie, the county borders Erie County, Pennsylvania, on the west; Warren County, Pennsylvania, on the south; and Erie and Cattaraugus Counties, New York, on the east. Fixing or determining the county's western boundary was a bit difficult as New York State's original charter established the western boundary along the south shore of Lake Erie to the forty-second degree of latitude on a line extended from Lake Ontario's western extremity.

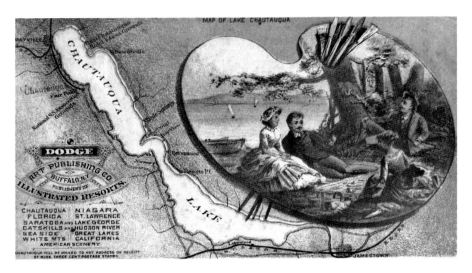

This nineteenth-century print celebrates the romantic mystique of Chautauqua Lake. The lake became an early center for rest, recreation and romance. *Courtesy of Chautauqua County Historical Society. Prepared by Niles Dening.*

This historic sign marking the old French Portage is located at the intersection of South Portage and Gale Street in Westfield, New York. *Photograph by Niles Dening.*

Determining Lake Ontario's western extremity required having New York and Pennsylvania come to an agreement on whether the "western extremity of Lake Ontario" included Burlington Bay or began at the peninsula dividing Burlington Bay from Lake Ontario. Andrew Ellicott from Pennsylvania and Frederick Saxton of New York were dispatched to run a survey and determine the boundary. They decided that the peninsula was the proper point and drew their line accordingly. This left a triangular tract of land called the Erie Triangle; it wasn't included in the charter of either state.

The Erie Triangle therefore reverted to the federal government. Pennsylvania finally purchased the triangle in 1791 to gain access to Lake Erie. Native Americans were living in the region at the time—Erie Indians were the original inhabitants. They were living along the southern shore of Lake Erie when French voyageurs first ventured into the region during the early 1600s. After a series of wars between the Erie nation and the Iroquois Confederacy, the Senecas seized the region in the mid-1600s. Senecas were living in the region when René-Robert Cavelier, Sieur de La Salle, sailed his *Griffin* along the southern shore of Lake Erie in 1679.

La Salle was the one who suggested linking the French settlements in Quebec and on the Great Lakes with the French settlements in Louisiana. He believed that this could be done by building a link of forts and settlements along the Great Lakes and down the Allegheny and Ohio Rivers. This would secure the interior of the North America. Marquis du Quesne decided to implement the concept when he became governor of Canada in 1752. The French started by building a fort at Niagara and creating a landing at Barcelona Harbor in what is now Chautauqua County.

The French troops then cut an eight-mile portage road over the ridge to Chautauqua Lake in 1753. The French Portage Road passed through what is now Westfield, New York, and ended on the shore of Chautauqua Lake in what is now Mayville.

Once they reached Chautauqua Lake, they had access to the Allegheny River, Ohio River and the Mississippi River all the way to New Orleans. The French later found a shorter portage between Presque Isle (now Erie, Pennsylvania) and French Creek, where they built Fort Le Boeuf. However, some French and later English travelers continued using the Chautauqua Portage up to, during and after the French and Indian War.

Chautauqua Lake is the dominant geographic feature of Chautauqua County; it cuts a seventeen-mile diagonal swath across the county. Although the lake is only about eight miles from Lake Erie, it is about seven hundred feet higher than Lake Erie. Chautauqua Lake varies in

This early photo of steamboats gathered at Mayville Landing on the upper end of Chautauqua Lake illustrates the commerce on the lake. Steamboats plied the lake during the late nineteenth and early twentieth centuries before being replaced by railroads, trolleys and automobiles. *Courtesy of Chautauqua County Historical Society. Prepared by Niles Dening.*

The McClurg Mansion, built in 1818, houses the Chautauqua County Historical Society. The museum has displays on three floors and contains a research library. The building was placed on the National Register of Historic Places in 1984. *Photograph by Hoffman.*

width from about one thousand feet wide in the middle to about two miles wide in three places. Steamboats once dominated the lake, but they've been replaced by pleasure craft.

Although Chautauqua County borders Lake Erie, most of the county is in the Allegheny Basin, being drained by the Allegheny River. People from Pennsylvania and New England started moving into the county after the American Revolution ended. Most agree that the first significant settlement in the county occurred at the "Cross Roads," now the village of Westfield.

James McClurg, from Pennsylvania, purchased land in 1801. He came up the Allegheny River, crossed Chautauqua Lake and got to Westfield in 1810. He built his fourteen-room Federal-style mansion in 1818. His mansion, the McClurg Mansion, now houses the Chautauqua County Historical Society. Founded in 1883, the historical society is one of New York State's oldest.

Several families from New England moved to Chautauqua County in the early 1800s. Most came in on a road cut out by the Connecticut Land Company, leading to Connecticut's Western Reserve, now northern Ohio. The road started at Buffalo Creek (present-day Buffalo, New York). The road ran along the southern shore of Lake Erie. It was cut out a distance back from the lake to avoid the bogs, swamps and mosquitos along the lakeshore. The road is still there. It's now called U.S. Route 20.

Some newcomers settled on the flats along Canadaway Creek in northern Chautauqua County. Eliphat Burnham came from Hartford, Connecticut, in 1805. Zattu Cushing, from Plymouth, Massachusetts, arrived later that year. Bennington, Vermont's Thomas Abell opened the first inn. Leverett Barker came from Branford, Connecticut. He opened a tannery in 1809.

Dozens and then hundreds of settlers moved to Chautauqua County after the War of 1812 ended. Many came on the new "Chautauqua road." What's left of that highway is now called the "Old Chautauqua Road." It was created during the War of 1812 to avoid the conflict taking place along the Niagara River. After the road was bridged in 1814, it became the route of choice for most settlers coming from the east. So many families from Vermont settled on and along the road in the central part of Chautauqua County that they created a community, calling it Vermont Village. It's now called Gerry, named after Elbridge Gerry, fifth vice president of the United States.

Jamestown, at the southern end of Chautauqua Lake, was settled by Matthew Prendergast in 1807. He came from Pawling, New York. Other settlers joined him, coming from Rhode Island, New Hampshire and Pennsylvania. They built a dam, sawmills and a gristmill at the outlet of Chautauqua Lake. Jacob Fenton, from Mansfield, Connecticut, opened a pottery in 1813. Royal Keyes, a millwright from New Fane, Vermont, arrived in 1816. Business thrived at the Rapids.

Dunkirk was settled on Lake Erie in 1810. Solomon Chadwick came from Weston, Massachusetts. Elisha Jenkins, Isaiah Townsend, John Townsend and DeWitt Clinton came from Albany, New York, and purchased one thousand acres of land around the harbor in 1816. They wanted to open a port on Lake Erie, so they constructed a pier and put up several buildings. They sold the property to Walter Smith in 1825.

People kept coming. There were 12,568 people living in Chautauqua County in 1820. This increased to 34,671 in 1830 and 47,975 in 1850. About 134,000 people now live in the county.

PART III
EARLY ACTIVITIES IN NATURAL GAS

WILLIAM HART AND NATURAL GAS: FREDONIA, CHAUTAUQUA COUNTY, NEW YORK

People have known about natural gas for a long time. Greeks, Persians, Romans and American Indians all noted and reported flames coming from burning springs. These "mysterious fires" were sometimes ignited by lightning; they drew worshiping pilgrims to be mesmerized by the miracle flames. Greeks believed that the mysterious fire on Mount Parnassus marked the presence of supernatural forces. They built a temple for the Oracle at Delphi.

Ruins of an old Parsee temple are found on the Caspian Sea. The structure was dedicated to the deity of fire. Stories of flaming rocks and bushes are found in the Torah and Bible. Plutarch described eternal fires burning in present-day Iraq.

However, no one seems to have tried using natural gas as a fuel to generate energy or create light until the early 1820s. This happened in Western New York. It took place in a settlement called Canadaway, located on flats along Canadaway Creek. The name was derived from the Seneca word *Ga-na-da-wa-o*. It meant "running through the hemlocks." Early settlers kept the name but muddled the pronunciation.

The name didn't last. When village leaders incorporated their settlement as a village in 1829, they wanted a more impressive, "classical," Latin-sounding name, so they used the word that Samuel Latham Mitchell suggested as a fitting name for the United States. They incorporated the village as "Fredonia," and so it has remained. Mitchell, according to the legend, created his unique word by taking the first syllable of the English word "freedom" and coupling it with a Latin-sounding inflection or ending.

William Austin Hart came to Fredonia in 1819, arriving "with his trade, rifle and pack—his fortune—and commenced business in a small way," according to the account in his obituary. William Hart was the son of Aaron Hart and Annie Austin Hart. He was born in Barkhamsted, Connecticut, on January 13, 1797, the first of seven children.

William Hart mastered the gunsmith's trade by working as an apprentice in Hartford, Connecticut. When he moved to Fredonia, he opened a gun shop on the second floor of a large U-shaped building on West Main Street. Getting established, he married Mary Ann Summerton on January 22, 1821. Sometime during the early 1820s, William took an interest in the gas bubbles popping to the surface of Canadaway Creek. Some say that this occurred in 1821; Joseph Holmes and others insist that it took place in July 1825. Joseph Holmes wrote a lengthy letter about the event, several years later. He claimed that it occurred when he was boarding with the Hart family while he was attending the Fredonia Academy.

Joseph Holmes described the occasion by saying that William Hart drilled a hole in the bottom of his wife's washtub in July 1825. Taking the washtub, Hart turned it over and put a rifle barrel in the hole. Then Mr. Hart smoothed out an area in the creek and placed the washtub over bubbles rising to the surface. After a considerable amount of gas had accumulated in the tub, Mr. Hart lit a candle. Placing the candle against the end of the rifle barrel, he lit the gas. He got a continuous flame.

Ms. Lois Barris provided an account of William Hart's follow-up activities in her account entitled *Fredonia Gas Light and Water Works Company*:

> *Mr. Hart made three attempts at drilling in the creek—he left a broken drill in one shallow hole and abandoned a second site at a depth of forty feet because of the small volume of gas found. In his third attempt, Mr. Hart got a good flow of gas at seventy feet. He constructed a crude gasometer, covering it with a rough shed and proceeded to pipe and market the first natural gas sold in this country.*

This is a contemporary photo of Canadaway Creek. William Hart obtained natural gas from this creek in the early nineteenth century. He sold it to nearby businesses and neighbors. *Photograph by Hoffman.*

Mr. Hart's early gas customers paid $1.50 per year for each light. One gas light was claimed to yield the light of two good candles. The owner of the mill, on whose land the well was drilled, received two free gas lights for his office as royalty.

Early customers were connected by log pipes during the fall of 1825. More customers were added over the course of the following year as the log pipes were extended. The gas burner in Noah Whitcomb's shop was described as "made by pinching the end of the lead pipe together so as to spread the flame."

The *Fredonia Censor* on August 31, 1825, wondered:

What village can compare with Fredonia? There is now in this village two stores, two shops, a mill, and a hotel that are every evening lighted up with as brilliant gas lights as are to be found in any city in this or in any other country. The hydrogen gas, or inflammable air, which produces these lights, was procured by drilling a hole, several feet, into the rock which composes the bed of the creek passing through this village.

Wooden pump-logs were used until William Hart started replacing these with lead and tin piping. The Abell House, a hotel, used natural gas for lighting and, some say, cooking. Small tin tubes were crimped into a "fishtail" shape to create gaslights. It was most impressing that the burning gas did not emit an odor.

A village resident observed, when she looked down the valley of the creek, that "windows were bright at night from the light of the wonderful gas spring." Thirty-six gaslights burned in the village by November 1826. William Hart was a businessman. Noting a curiosity, he created a product, sold it and made money.

The *Fredonia Censor* printed an update on William Hart's gasworks on November 30, 1825:

> As this is undoubtedly the first attempt which has ever been made to apply Natural Gas to so extensive and useful a purpose, we shall give a brief description of the manner in which it is effected in this village. A hole was drilled 27 feet into a slaty rock on the margins of the creek, from whence the gas issues—from this it is conducted about 25 feet in lead pipes and discharged into a vat 6 by 8, and 4 feet deep, excavated out of solid rock, and which is filled with water.
>
> Over this vat is suspended the gasometer, which is constructed of sheet-iron and will hold upwards of 1,200 gallons, in such a manner, that when it is sufficiently filled it rises within two or three inches of the top of the water, when this over-plus of the Gas escapes under its edge, and is drawn off for use, this gasometer again settles down into the vat.
>
> A wall and an arch of substantial masonry are crested over the gasometer, with doors for the admission of the curious. The Gas being conducted into buildings by lead pipes is then conveyed to any part of the building by means of tubes at the end of which is a burner…as no failure of this gas is ever anticipated, the works have been constructed to last for ages.

Being a master mechanic, William Hart developed other products. He got the country's first patent for a percussion-lock firearm. The percussion lock replaced the older flintlock. William Hart's patent is dated February 20, 1827. It was signed by President John Quincy Adams, Secretary of State Henry Clay and Attorney General William Wirt. The United States commissioner of patents even sent William Hart a congratulatory letter on February 23, 1827, saying, "I am pleased to find so much genius accompanied with so much morality."

Becoming older, William Hart took up gardening, creating a nursery on land he owned along Canadaway Creek. Expanding his nursery from today's Hart Street to the creek's edge, he turned his nursery into Hart's Pleasure Garden. He planted elaborate flower beds, put up arbors and laid out brick walkways.

The *Fredonia Censor* noted, "Walks are in charming order and are illuminated by gas lights and torches for evening rambles." In the announcement marking the grand opening of his 1837 season, William Hart assured prospective customers: "The Pleasure Gardens will be illuminated for a grand gala and pyrotechnic entertainment will be given worth the extent of patronage that may be offered."

His season ticket, entitling one to the privilege of both the Pleasure Garden and Baths, was five dollars (or three dollars if separate). Food bars were an important feature of Hart's Pleasure Garden. Tables were loaded with refreshments, "in all of their seasonal varieties." Bathhouses, heated by natural gas, opened on May 31. The bathhouses were open every Saturday until the hot season began. Then they were open every day of the week except Sunday. They provided hot, cold and shower baths.

Mr. Hart's special July 4 celebration was elaborate. He opened the Pleasure Garden for the entire day. Food bars were stocked with a variety of fruits, baked goods and other luxuries. Admittance to the day's festivities cost a shilling, about twenty-five cents; another shilling, about fifty cents total, included fireworks.

The highlight of the day was the afternoon balloon ascension. The balloon, the largest ever seen, was inflated with natural gas from Mr. Hart's gasworks. Fire rockets capped off the evening entertainment. The *Dunkirk Beacon* reported, "The day and entertainment were as successful as Mr. Hart hoped."

The Hart family left Fredonia in 1838, moving to Buffalo, New York. William Hart became a successful and prosperous businessman in the city. The 1850 census records show William Hart residing in Buffalo's Fifth Ward. Working with the Buffalo Gas Light Company, he listed his occupation as merchant.

The Buffalo Gas Light Company was one of the country's first gas-manufacturing companies. It was housed in a large, church-like, stone building located at the foot of Genesee Street on Buffalo's waterfront. Coal was unloaded from boats and burned in sealed ovens, creating illuminating gas. This gas was piped to the surrounding area and used for lighting streetlamps, homes and some businesses. There were 179 streetlamps.

Above: This shaded field, running along Canadaway Creek, was once the site of William Hart's Nursery and Pleasure Garden. *Photograph by Hoffman.*

Left: Hart Street, a short street in Fredonia, honors William Austin Hart. He lived at 50 Forrest Street in the early nineteenth century and owned the land behind his house, along Canadaway Creek. *Photograph by Hoffman.*

The 1860 census reported William Hart, his family and his married son's family living in Buffalo's Tenth Ward, a more upscale community. This time, William and his son each listed his occupation as "gas furnisher."

William Hart rushed to Pennsylvania with Preston Barmore during the early 1860s to work on the oil fields. William died in Buffalo on August 9, 1865; he left a legacy of light and energy.

Fredonia, New York, recognized William Hart's contribution by naming a street after him. Hart Street is a short street that runs from Forest Place to Canadaway Creek. The street borders a field, once the site of Hart's Pleasure Garden.

GENERAL LAFAYETTE VISITS FREDONIA: CHAUTAUQUA COUNTY, NEW YORK

General Lafayette, hero of the American Revolution, visited the United States for the last time in 1824. He landed at New York in August and took a tour of the country, going west to the Mississippi and coming back through the northern states. He was received with great enthusiasm; high honors were paid to him everywhere he went. Getting word that he was coming to New York State to view Niagara Falls, a number of local dignitaries, all gentlemen, gathered in Westfield to plan an appropriate reception.

The Honorable William Peacock, the resident Holland Land Company subagent, provided a superb carriage for the conveyance of the general. The gentlemen waited for General Lafayette at the state line and greeted him when he appeared. After he changed carriages, the gentlemen escorted him to Westfield. When they reached Westfield shortly after sunset, a large crowd was waiting to greet General Lafayette. There was a lengthy reception, several speeches and a grand artillery salute. General Lafayette got back into the carriage a little after 10:00 p.m. to continue on his way.

He was escorted to Fredonia by a local militia company. They got to Fredonia a little after 2:00 a.m. on Saturday, June 4, 1825. Even so, Captain Whitcomb's rifle rangers, Captain Brown's artillery company and a detachment from the 169th militia regiment waited at the top of West Hill to greet him. Captain Brown's artillery company announced his arrival by firing a grand thirteen-gun salute.

The Barker Library and Museum is located on Day Street in Fredonia, New York. The museum has an archive, two research libraries, exhibitions and a children's museum. General Leverett Barker constructed the original building in 1821. *Photograph by Hoffman.*

Then the militia formed up around the carriage to escort General Lafayette down West Hill, over the log bridge and up Main Street. The street was lit with candles and lamps. Controversy persists in Fredonia, to this day, regarding the lamps. Were they gas or oil?

Hamilton Child's *County Gazetteer and Business Directory*, published in 1873, claims gas. A statement in the *Gazetteer*, in the form of a footnote, read:

> *The use of natural gas at Fredonia was begun in 1821, when experiments were made to determine its illuminating value, and it was introduced into a few of the public places, among which was the hotel...and which was thus illuminated when La Fayette passed through the village.*

Joseph Holmes, in his later letter, disagreed. He claimed that gaslights were first lit in Fredonia in August 1825. Lafayette was in Fredonia two months earlier, June 1825. Nevertheless, Fredonia residents were up, about and ready for the occasion. They gathered in front of the Abel House to greet General Lafayette. Ladies, Revolutionary War veterans and several other citizens formed two long reception lines that led to an elevated stage erected in front of the hotel. General Lafayette and his entourage got out of the carriage, lit by the light of flaming lamps.

General Lafayette's secretary described their Fredonia reception in his 1825 notebook: "I shall never forget the magical effect produced at Fredonia... Our eyes were dazzled by the glare of a thousand lights, suspended to the houses and trees that surrounded us."

A memento of the occasion remains. General Leverett Barker illuminated his mansion with candles in each window. One part of a sash was scorched, and Barker never allowed it to be painted over. It's still there—in the Barker Library. A brass marker was placed next to the window in 1900 to commemorate the reminder of Lafayette's visit.

As time went on, other travelers noted streets and public places lit by natural gas in Fredonia. Reports traveled by letters, newspaper articles and stagecoach. Professor Benjamin Silliman at Yale University, editor of the *American Journal of Science and Arts*, heard about it. He published an account of natural gas being used for lighting at Fredonia in his January 1830 issue. Sir David Brewster, a Scottish physicist, republished the report in his *Edinburgh Journal of Science*.

Baron Alexander Humboldt is said to have commented on the discovery and use of natural gas at Fredonia to be "the eighth wonder of the world."

BARCELONA LIGHTHOUSE

The United States Congress declared Portland Harbor, a Chautauqua County port on Lake Erie, an official "port of entry" in 1827. The following year, Chautauqua County's congressman, Daniel Garnsey, got Congress to set aside $5,000 to purchase land and build a lighthouse at the harbor. Local folklore has it that Congressman Garnsey got the bill through Congress by getting Henry Clay's support.

Henry Clay, a congressman from Kentucky, had served several terms as Speaker of the House. As Henry had urgings for the presidency, he wanted to get his name out and about. Congressman Garnsey secured Henry's backing for the lighthouse project by convincing investors who were building a new steamboat at Black Rock, on the Niagara River, to name their boat the *Henry Clay*.

This old photo of the Barcelona Lighthouse was taken sometime in the late 1800s. Barcelona, New York, was once an important commercial fishing port. Sailing vessels brought back whitefish, herring, bass and fifty-pound sturgeon. *Courtesy of Chautauqua County Historical Society. Prepared by Niles Dening.*

It became the fourth steamboat on Lake Erie. The *Henry Clay* was described as "being worthy of the name of the great Western orator and statesman." It was powered by a new sixty-horsepower steam engine; its cabins were described as "elegant" and "expensively fitted up."

Congressman Garnsey came from Columbia County in eastern New York. He was an attorney, served as an army officer during the War of 1812 and moved to Chadwick Bay, now Dunkirk, in 1816. Daniel Garnsey held a number of local offices before being elected to the United States Congress in 1824. He served two terms, four years, from 1825 until 1829. After securing the lighthouse appropriation, he worked with Judge William Peacock to pick an appropriate site. They selected a location overlooking Barcelona Harbor.

Judge Peacock was then appointed construction supervisor for the project. Judge Peacock hired Judge Thomas Campbell, from Westfield, to build the lighthouse. The construction contract stipulated that a keeper's cottage, water well and outhouse were to be built along with the lighthouse. The site was purchased from the Holland Land Company for $50.00. Construction costs came to $3,456.78. Work was completed by June 1829.

Judge Campbell knew that natural gas had been used to light commercial buildings in nearby Fredonia. Believing that natural gas could be used to light the new lighthouse, he formed a partnership with Charles C. Tupper. Charles owned of a parcel of land near the site known to contain a "burning spring." They issued a contract on January 1, 1831, hiring William Hart, Fredonia gunsmith and master mechanic, to provide the lighthouse with natural gas "at all times and seasons" and to keep the apparatus and fixtures in repair at an annual cost of $213.

William Hart and others went to work securing natural gas from the "burning spring" in Tupper Creek and conveying it to the lighthouse. Workmen dug a hole in the rock at the place in the creek where the largest quantity of gas was detected. Then they enlarged the hole by digging a collection well about forty feet in diameter and three feet deep. That being done, they constructed a solid masonry cone over the well. The structure was tight enough to contain the natural gas collected in the cone and, at the same time, exclude water flowing around the base of the cone.

The workmen put a lead pipe in the bottom of the cone. The pipe was bent down at the end, aimed toward the bottom of the well. Then they ran lead pipe along the bed of the creek to a termination point below the dam. From that point, gas was conveyed to the lighthouse inside wooden pipe created from drilled pump logs. The builders sealed the joints and buried the logs in the ground. They ran the log pipe about 380 feet to the base of the lighthouse.

They used iron pipe to convey the natural gas to the top of the lighthouse. William Hart designed and built a stand of lights to receive and burn the gas. His light stand consisted of thirteen horizontal arms extended like the radii of a semicircle. A brass pipe burner was attached to the end of each arm. A stopcock regulated the flow of gas consumed by each burner. Each burner was provided with a large and suitable reflector. There were two tiers of lamps, seven in the lower tier and six in the upper tier. Burners were interspaced so that when viewed from the lake during the night, the lighthouse provided a single, complete, constant, unwavering light.

It worked because the lighthouse happened to be higher than the source of the gas. Joshua Lane got the lighthouse keeper's job on May 27, 1829. Contemporary accounts described Joshua as a "deaf, superannuated clergyman, having numerous female dependents." Joshua's salary was set at

$350 per annum. The light was lit on July 5, 1830, becoming the world's first lighthouse to be lit by natural gas. The *Fredonia Censor* commented:

> *The Lighthouse at Portland Harbor in the County of Chautauqua and State of New York is now illuminated, in the most splendid style, by natural carbureted hydrogen gas. Ever since the first settlement of the country about Portland, it has been known that an inflammable gas constantly issued from the fissures of a rock, near the harbor, in such quantity as to be easily set on fire by applying a flame to it.*

Three long docks and four warehouses were in use at Portland Harbor by 1831. The federal government constructed a larger, longer pier and breakwater in 1844 at a cost of $56,000. Then the Great Gale of 1844 struck without warning. The wind started blowing on the evening of October 18, 1844. Huge waves washed over Barcelona, Dunkirk and Buffalo Harbors that evening, striking after midnight. The gale destroyed shipping, warehouses and docks. Dozens of people died. However, local officials decided to rebuild the port at Barcelona.

They did and lake commerce recovered, but then the Buffalo and State Line Railroad came through nearby Westfield, New York, in 1852. The railroad siphoned off most of the trade formerly carried by ships on the lake. Three years later, in 1855, the federal Lighthouse Commission recommended to the United States Congress that the Barcelona Lighthouse be closed, as it was no longer needed.

A condition of the original deed stated that the property would pass back to the original owner at such time as it was no longer needed. Having served as a beacon of progress for about a quarter of a century, the lighthouse reverted to private ownership. Even so, the old Barcelona Lighthouse is still standing. It dominates the harbor.

New York State Geological Survey: Dr. Beck's Report on Natural Gas

Governor DeWitt Clinton asked Amos Eaton, a geologist and cofounder of New York State's Rensselaer Polytechnic Institute, to give a series of lectures before the New York State legislature in 1818. Professor Eaton did so, stressing the need for a systematic natural resource survey of New York State. He started the process by completing surveys in Albany and Rensselaer Counties and along the Erie Canal. The state legislature followed up a few years later, in 1834, by directing Secretary of State John Dix to come up with a plan to complete a four-year survey of the state's resources. The project was funded with four annual $26,000 appropriations.

Two years later, in 1836, Governor William Marcy appointed Dr. Lewis Caleb Beck to conduct a geologic survey of the entire state. Dr. Beck was a graduate of Union College who had studied and practiced medicine for a while before becoming a professor at Rensselaer Polytechnic Institute. He later moved on to become a chemistry and pharmacy professor at Albany Medical College.

Dr. Beck traveled more than eight thousand miles conducting his survey, visiting most of the state from year to year. He collected, in his words, "many suites of specimens for the General Cabinet…and devoted the rest of my time to arranging the materials collected, and to the analysis of such rare and useful products as seemed worthy of examination." Dr. Beck's personal collection, and those of his colleagues working on the survey, provided the foundation to the State Cabinet of Natural History, later to become the New York State Museum.

Visiting Western New York, Dr. Beck noted Fredonia's natural gas springs. He published a report of his findings to the New York State legislature, concluding:

> *The gas springs seem to have their origin in the strata of slate which form the bed of the stream and which are everywhere met with in this vicinity a short distance from the surface of the earth. This slate has a bluish color, and some of the layers are exceedingly fragile, requiring only a few years exposure to be completely converted into a clayey soil. The lower strata, however, resist atmospheric agencies and are sometimes used as building material.*
>
> *When recently broken, the slate always emits a strong bituminous odor, and it frequently contains thin seams of a substance resembling bituminous coal. Most commonly, however, this bituminous matter occurs in patches, having more the appearance of detached vegetable impressions than of regular stratum.*
>
> *A fact I observed at Fredonia, in my opinion proves that there is at some distance below the surface, a vast reservoir of gas, the evolution of which is prevented by the pressure upon it. The fact to which I refer is that when the water in the creek is low, bubbles of gas are often observed, which*

disappear entirely when the water has risen, after a rain. And again, gas may be obtained at almost any part of the bank by boring to a depth of twenty or thirty feet.

So common, indeed, is this occurrence that many of the wells in the village of Fredonia are strongly charged with gas. It may also be added, that there are frequently to be observed in this vicinity disruptions in the strata of slate, which have probably been caused by some expansive force exerted from beneath.

Bubbles of gas are here and there seen rising through the water in this creek for nearly three quarters of a mile below the village. But the largest quantity is evolved at the later point. It was not possible for me, with any apparatus which I could command, to determine the amount of gas given out at this place in a given time; but bubbles rise with great rapidity from an area of more than twenty feet square, and I would probably be warranted in assessing that it five or six times greater than that which is currently being obtained at the village.

Although Dr. Beck published a number of books and papers on botany and chemistry, he is most noted for his comprehensive report on the mineralogy on New York State. His *Natural History of the State of New York* was published in 1842.

Fredonia Gas Lights and Waterworks Company: First Natural Gas Company

Some Fredonia residents complained about their natural gaslights; they claimed that lamps flickered. Others believed that their lamps did not burn as brightly as those being lit by Mr. Hart's manufactured gas in nearby Buffalo. By the 1850s, civic-minded Fredonians had decided to take action. They petitioned the New York State legislature for a charter to organize a corporation to supply their village with a better, more efficient gas supply and a supply of pure water to meet the needs of the growing community. A bill to accomplish these tasks was introduced in the New York State Assembly on April 4, 1856, but it got held up in committee.

Reverend Isaac George, the New York State assemblyman from Fredonia, vigorously pursued the reintroduction of an act required to establish the

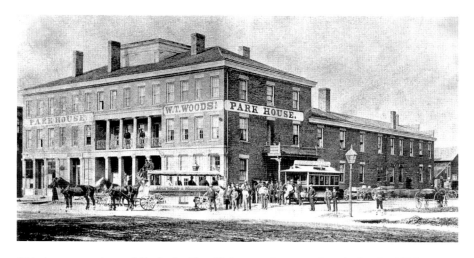

This downtown photo of Fredonia, New York, was taken sometime during the 1850s. *Courtesy of D.R. Barker Historical Museum. Prepared by Niles Dening.*

company. He labeled his new bill *An Act to Incorporate the Fredonia Gas Light Company*. When his bill cleared the committee on March 2, it had become an *Act to Incorporate the Fredonia Gas Light and Water Company*. On motion by Assemblyman George on March 16, his bill, now no. 385 and titled an *Act to Incorporate the Fredonia Gas light and Water Works Company*, passed the New York State Assembly. It was sent to the New York State Senate.

The bill came back from the Senate, with several amendments, on March 25. The Assembly discussed these amendments. Mr. George answered all questions, vigorously defending his bill. On Monday, April 13, 1857, the New York State Senate unanimously passed the act to incorporate with amendments, and on Tuesday, April 14, the State Assembly finally passed an *Act to Incorporate the Fredonia Gas Light and Water Works Company*.

With Governor Clark's approval, the Fredonia Gas Works and Water Company became New York State's first energy company. The company stated its objective in its charter: "By boring down to slate rock and sinking to sufficient depth to penetrate the manufactories of nature; to thus collect from her laboratories the natural gas and purify it; to furnish the citizens with good, cheap light."

Incorporators and first directors of the energy company were Preston Barmore, Rosell Greene, Orson Stiles, Thomas Glisan, Isaac George, Stephen Palmer and George Lewis. The directors announced the organization of their company in July with $10,000 in capital stock, to be sold at $25 per share. Then they proceeded to make a survey of needs in establishing a water and gas department. Officers were elected: Orson Stiles as president, Preston Barmore as secretary and Thomas Glisan as treasurer.

Willard McKinstry, editor of the *Fredonia Censor*, encouraged the public to cooperate with the new enterprise: "The company thus being fully organized, there is no good reason why the work should not progress rapidly if our citizens back up the enterprise with liberal subscription as it deserves."

The following week, the *Censor* announced that $2,000 had been subscribed in the first two days of the stock sale. Within a month, it was reported that exploration made on the land owned by T.W. Glisan of Green Street resulted in the development of springs of cold water sufficient to supply the citizens of Fredonia three times over. Water being obtained, Preston Barmore got to work trying to find sufficient quantities of natural gas.

Preston Barmore Discovers Fracking: Foreshadowing the Future

Preston Barmore was born in Forestville, New York, in 1831. He was the oldest son of Lewis and Diana Trask Barmore. The Barmore family came to Chautauqua County from Rhode Island, initially settling in Hamlet, a community on Conewango Creek in the town of Villanova. The family then relocated to Forestville in the town of Hanover. Lewis Barmore opened a cabinetmaking shop in Forestville and ran an undertaking business.

Growing up in Forestville, Preston enrolled at the Fredonia Academy in 1847 when he was sixteen years old. He attended classes at the academy until he completed the program in 1851. Then Preston worked at several jobs in and around Fredonia until he convinced others, more affluent Fredonians, to invest their money by creating a gas company. They did this and then petitioned the legislature.

The Fredonia Gas Light and Water Works Company became incorporated in April 1857. Preston Barmore was one of the first directors, along with Rosell Greene, Orson Stiles, Thomas Glisan, Isaac George, Stephen Palmer and George Lewis. Orson Stiles served as president. Preston Barmore became secretary and Thomas Glisan was elected treasurer. The directors selected S. Miner Newton to be their engineer.

Within a month, Newton had found springs of pure water on Thomas Glisan's property near Prospect Street. The springs produced enough water to supply the village three times over. Meanwhile, Preston Barmore got to work. Having read Dr. Beck's report while a student at the Fredonia Academy, Preston a fairly good idea where to drill.

Walking down the Canadaway Creek, Preston came upon a gas spring about one mile north of the village. Using energy company funds, he purchased one-fourth of an acre of land. Using a spring pole device, he started kicking down a well.

His spring pole device was a simple thing, relatively easy to make. It consisted of a long pole, a weight to anchor the butt end of the pole, a fulcrum, some rope and a few iron drilling tools. Preston probably used hemlock for his spring pole. Hemlock was available, springy and didn't crack easily. His fulcrum was probably a forked oak post, with the spring pole resting in the fork. A length of rope was looped around the spring pole, creating stirrups.

Drilling required strong legs, as the driller kicked and released the stirrups, driving the drill deeper into the ground with each kick, drilling a four-inch borehole. Reciprocating action helped drive the drill deeper. A high tripod of poles was erected over the borehole, and a pulley was hung from the tripod. This helped the driller pull out the drill string when he needed to sharpen or change his tools.

Preston got gas, but the flow wasn't satisfactory, so he drilled another well. By mid-December, Preston's second borehole was 127 feet deep. Once again, he got gas, but it wasn't enough. Then Preston came up with the idea of "cracking or breaking" the subsurface rock, creating artificial fractures to increase production. Today, it's called "fracking."

"Fracking a well" consists of fracturing the rock to allow gas and oil to flow, making it easier to get natural gas from the earth. Preston "fracked" his well by dropping a metal canister packed with eight pounds of gunpowder down the four-inch-diameter shaft of his well to the bottom. Then Preston pushed a hollow tin tube down the shaft. When that was done, he heated a hunk of iron and dropped the red-hot piece of iron down the shaft. It ignited the gunpowder, expelling water and a plentiful flow of gas—enough natural gas to supply 1,200 lights or burners.

The *Fredonia Censor* reported that Mr. Barmore embarked on the project with "energy and perseverance" and hoped that his experiments would meet with a more substantial recompense, as his geological theories had "thus far been tested entirely at his individual expense."

The *Fredonia Advertiser* printed a progress report on December 18, 1857:

> *The Fredonia Gas Light and Waterworks Company chartered and organized for the purpose of supplying the village with water and gas have succeeded in laying down for their water works a "Street Main" from the reservoir to the corner of Main and Mechanic Street…it is expected they will resume operation in the spring for completing both the water and gas communications.*

A network of lead pipes was entrenched throughout the village. Main lead pipes were four inches in diameter, and branch pipes were two or three inches in diameter; pipe joints were packed and sealed with lead. The village contracted the gas company to install six lampposts on the village streets. By June 1859, most of the stores and businesses in Fredonia and several of the street corners were lit by gas lanterns and lights.

The *Cleveland Herald* reported:

> *Another thing which the citizens of Fredonia flatter themselves (and we think not without reason for doing so) is the project of lighting the streets with natural gas…with these projects consummated, Fredonia, with its already many allurements, will certainly become one of the most inviting residences in the country.*

Preston Barmore and his workmen followed up by digging a natural gas collection chamber in the bed of Canadaway Creek. Their collection chamber was about six feet in diameter at the top, thirty feet deep and twenty feet in diameter at the bottom. Two gas-producing wells were then drilled into the rock.

Workers were reminded of the nature of natural gas when, sometime in March, one of the workmen, Burt Flanders, was injured. Burt tried to light his pipe while standing in the cavity, above the wells. The resulting blast "exceeded in volume all anticipation and such was the force of it, Mr. Flanders was thrown bodily out of the pit and made a very judicious exit though attended with serious consequences."

A few weeks later, the *Censor* reported, "The effects of the explosion are still visible on Mr. Flanders' face and hands." In August, the *Censor* announced that "Mr. Preston Barmore has completed the arch over his gas well and collection chamber, the excavation of which has occupied eleven months." The account further noted that the amount of gas derived from the pit would soon be determined by gas meters under the direction of Mr. William A. Hart of the Buffalo Gas Works. This was the same William A. Hart who had drilled the first natural gas well in Fredonia.

A *Fredonia Censor* reporter later visited the operation at Barmore's gasworks, noting:

> *The well is arched over and two pumps are in operation for clearing it of water, cost $150. Measure of the gas was made under the supervision of William A. Hart, Esq. with one of Samuel Dawn's Dry Meters, owned by the Buffalo Gas Company. The total amount being a little over 9,600 cubic feet, the well can supply 1,200 lights or burners. Results exceed the most sanguine expectations of Mr. Barmore.*

The Fredonia Gas Lights and Water Works Company went through reorganization in November 1858. The directors split the company into two corporations, the Fredonia Water Company and the Fredonia Natural Gas Company. Mr. H. Pemberton, who purchased half interest in the gas company, became president of the Gas Company. Elias Forbes was named treasurer and Preston Barmore secretary. Mr. Forbes contributed "capital, energy, and business experience." Mr. Barmore's contribution was described as "great energy, persevering scientific and practical knowledge." The company continued laying pipe throughout the village.

That summer, in 1859, Alfred L. Drake struck oil near Titusville, Pennsylvania. It did not take long for Preston Barmore to shift his attention from natural gas to oil. In the fall of 1860, a reporter for the *Titusville Gazette* visited the site on Oil Creek where Preston Barmore's Empire State Rock Oil Works had several wells. Mr. Barmore was described as "commencing as though sure of ultimate success." Unlike most of the early oil wells that were drilled on flats along the creek, Mr. Barmore was drilling under a bluff, a distance back from the creek. He had a good well, producing fourteen barrels of oil per day and five hundred barrels over the course of six weeks.

Mr. Barmore's drilling machinery was described as being "in perfect order, wasting nothing." Natural gas derived from his oil wells was collected and stored in reservoir containers. It was then used as fuel to heat the boilers that produced the steam that powered the steam engines used for drilling. The greatest of the early oil fields was that on Buchanan Flats on Cherry Run. The place soon became a village called Rouseville, named after Henry Rouse. Mr. Rouse, a successful lumberman and merchant, grew up in Westfield, New York, about twelve miles from Fredonia.

Henry Rouse got in on the oil action early as he had the money to purchase leases on several of the most likely oil sites along Oil Creek. However, Mr. Rouse suffered the misfortune of being killed in an oil well explosion at Buchanan Flats on April 17, 1861. Preston Barmore was one of the owners of the well. Preston survived the explosion and, after they got the fire out, profited from the well.

Eight months later, an identical item appeared in both Fredonia newspapers, the *Censor* and *Advertiser*: "Preston Barmore died Wednesday night 4 December, 1861, one of the most enterprising and energetic young men of the county."

The interment record of the Forest Hill cemetery recorded Preston Barmore's cause of death as "inflammation of rheumatism, producing insanity." Aaron Barmore, Preston Barmore's brother, happened to be Preston's undertaker; Aaron kept the records. Others claimed alcoholism, but perhaps it was a combination.

Commercial Use of Natural Gas: From 1856 to 1886

Seeing what was happening in Fredonia and at Barcelona Harbor, other communities started exploring the possibility of developing their own natural gas resources. Nearby Erie, Pennsylvania, chartered the Erie Gas Company, with a capital of $60,000, on March 5, 1852. The company purchased ten lots for $10,000 and spent $50,000 on buildings, machinery and equipment. It drilled a deep well (1,200 feet), producing natural gas in 1853. It built a gas holder, forty-five feet in diameter and eighteen feet deep. Three and a half miles of pipe were laid in the streets. Gas was piped to the homes of the first thirty-six residents on August 22, 1853.

The gas company was supplying 150 residents by the end of the year, lighting six hundred burners. H. Jarecki & Company's Petroleum Brass Works in Erie started using natural gas in 1868 to light the shop and heat its boilers. The gas replaced eight to ten tons of coal every month.

The Erie Gas Company continued drilling wells. It had twenty-five producing natural gas wells by 1870. Natural gas was used to fuel a soap factory, a brewery and a seminary and was used to light homes. Natural gas illuminated the city. Proponents proclaimed, "There is no smoke, no dust, no ash, and nothing to do except turn a faucet to either shut it off or turn it on."

Natural gas wells were also drilled in nearby Knox County, Ohio. Gas from these wells was used to light and heat several nearby homes. Peter Noff started using this gas to manufacture lampblack in 1873.

Commercial gas wells were drilled in Meade County, Kentucky, starting in 1863. Gas was burned to evaporate brine for salt and later delivered to Louisville for heat and lighting. Natural gas was subsequently discovered in Wellsburg, West Virginia, in 1869. It was used to light the streets of several nearby communities.

For the most part, the early use of natural gas required being close to the source. Gas, being lighter than air, tended to escape. William Hart tried using pump logs in his unsuccessful effort to get gas to the Dunkirk Lighthouse and used them again in his successful venture at Barcelona. The first twenty-five-mile, long-distance natural gas pipeline was constructed in 1870. It connected a source of natural gas found at West Bloomfield, New York, with Rochester, New York.

As the builders used wooden pipe, they ran into a number of problems. First of all, they had to deal with a drop in elevation, eighteen feet to the mile. Natural gas, being lighter than air, had to be forced through the line, but the joints leaked. The builders tried wrapping the joints with tar-soaked blankets, but that didn't work. Wooden pipe was a problem. It was cumbersome and hard to make. Pine logs had to be selected and drilled with caution. A three- or four-inch-diameter hole had to be drilled through the center of each log, from one end to the other. Drilling had to be carefully done, without splitting or cracking the log. Logs could be laid, but they couldn't be sealed.

William Smith, a Pittsburgh iron molder, fashioned a solution by using a wrought-iron pipe. His National Pipe Works Company started producing iron tubing, casing and drive pipe. The pipe, created from lap-welded wrought iron, was manufactured in eighteen- and twenty-four-foot lengths. It was fitted with a tapered, coarse thread at one end and a collar, called a socket or coupling, at the other.

The first successful long-distance natural gas pipeline using Smith's iron pipe was laid in 1872. It was a two-inch line running five and half miles from the Newton's wildcat well to Titusville, Pennsylvania. William Newton, John Wheaton and John Tracy drilled the well with the hope of getting petroleum. They got natural gas. The *Titusville Herald* described the event:

The well was drilled. Being a wet hole, it was cased with casing. There was only a slight indication of gas, until the water was exhausted. Then an intense flow of gas issued from the casing, throwing a column of water several hundred feet into the air. Noise from the rushing gas scattered cows on nearby farms and the dismissal of students at a school nearly a mile away.

Newton, Wheaton and Tracy sold their well to a group of Titusville investors headed by Henry Hinckly and A.R. Williams. These new owners laid the pipeline and used a steam-powered compressor to convey the output of "the most powerful and voluminous gas well on record" to Titusville for lighting and heating purposes. Honey, glue, mucilage and glycerin were used to seal the joints, but some still leaked.

Solomon Dresser, the founder and president of S.R. Dresser Manufacturing Company, finally solved the leaking problem by creating and manufacturing a special coupling that could be used to seal joints. A seventeen-mile, six-inch pipeline designed to carry natural gas was laid in 1876 from Butler County to Etna, Pennsylvania. This gas was used to supply an ironworks. Olean, New York, brought natural gas across the state line in 1883 to light the streets of the city "from sunset to dawn for two years from January 1, 1884, at a cost of two dollars a year for each lamp so lighted."

Several, more productive, natural gas wells were then drilled in western Pennsylvania. Finding a market for all this natural gas was a problem; getting it to market was yet another problem. Standard Oil interests in the Buffalo Natural Gas Fuel Company came up with a solution. They constructed the "world's longest natural gas pipeline" in 1886. This line conveyed natural gas from Kane, Pennsylvania, to Buffalo, New York. Seven hundred McKean County natural gas wells were drawn on to supply the city of Buffalo with sufficient quantities of gas. The eighty-seven-mile, eight-inch wrought-iron gas pipeline was considered one of the world's great engineering marvels at the time.

Hundreds of hired men dug the trench with picks, shovels, horses and wagons. Men camped out, living in temporary settlements pitched along the right-of-way. They slept in large dormitory tents and ate in huge mess tents. Twelve-foot lengths of heavy wrought-iron pipe were transported by railroads as close to the construction sites as possible, and then scores of horse teams, rented from local farmers, hauled wagon loads of pipe to the trench.

Each length of pipe was lowered by hand; workers wielding gigantic pipe wrenches threaded and coupled the joints. The most highly skilled men on the job were the "stabbers." They had to line up the pipe and start the joint without crossing the threads. According to an observer, the pipeline ran "up mountains, down valleys, under rivers, and around towns. Working summer and winter, the workmen got the pipeline to Buffalo in time to heat homes and roast turkey for the Christmas holiday."

This first successful long-distance natural gas pipeline came across Chautauqua County and a corner of our farm on its way to Buffalo. A tin shack compressor building, housing a huge engine running on natural gas, sat on a ridge on the other side of the Walnut Creek Valley. Old-timers claimed that the engine ran for fifty years without missing a beat, but it was quiet as I grew up.

One day in the late 1940s, the gas company came along with bulldozers, cranes and workmen to dig up the old pipeline. It was replaced by a larger line one mile or so down the Creek Road. I believe it's still there.

PART IV

PETROLEUM

Early Ventures

FRENCH CREEK, AN EARLY WATERWAY: WESTERN NEW YORK TO NORTHWESTERN PENNSYLVANIA

Waterways were once the thoroughfares of Western New York. French Creek was an important waterway. The creek begins near Sherman, New York, a Chautauqua County village in the Allegheny Basin. Sherman happens to be located about twenty-six miles southwest of Fredonia, about thirteen miles south of Westfield.

French Creek emerges from a wetland near Chautauqua Lake. Flowing through southwestern Chautauqua County, the creek leaves New York, crossing into Pennsylvania near Findley Lake. Then the creek meanders through Erie, Mercer, Crawford and Venango Counties in Pennsylvania before flowing into the Allegheny River at Franklin, Pennsylvania.

The creek served as an important link between the Great Lakes and Ohio River. A French expedition explored the region in 1753. After building a fort at Presque Isle on Lake Erie, they chopped out a wagon road, widening a fifteen-mile Indian moccasin path that crossed a ridge between Lake Erie and French Creek. Their wagon road, the Venango Trail, connected Fort Presque Isle with Fort Le Boeuf on French Creek. The creek provided access to the Allegheny River. The Allegheny flowed

French Creek Preserve is located near Findley Lake, New York. It is open to the public throughout the year. A loop trail traverses several habitats. These include creek frontage, a wooded floodplain, open fields and a mixed hardwood forest. The Loop Trail can be hiked in about one hour. *Photograph by Hoffman.*

into the Ohio, and the Ohio flowed into the Mississippi. The Mississippi took them to New Orleans.

The French held the region, Western New York and Western Pennsylvania, for about one hundred years. After building Fort Le Boeuf, the French built another fort, Fort Venango, where French Creek joins the Allegheny. They

This photo of French Creek was taken south of the village of Sherman, New York. French Creek played an important role in Western New York and northwestern Pennsylvania history. It was once an important waterway. *Photograph by Hoffman.*

did this in 1754 as part of their plan to secure the interior of North America. Then the French started building another fort, Fort Duquesne, at the junction of the Allegheny and Monongahela Rivers.

When Governor Robert Dinwiddie of Virginia got word of this activity, he dispatched a young militia major, George Washington, to the site to deliver a message to the French. Major Washington went to Fort Duquesne, and then he continued up the Allegheny to Fort Venango.

Major Washington got to Fort Venango on December 4, 1854. His journal tells us that he conferred with French officers at the fort before going up French Creek to Fort Le Boeuf. When he got to Fort Le Boeuf, Washington spoke with the commander, Captain Jacques de Saint-Pierre. Washington delivered Governor Dinwiddie's note, part of which stated, "It becomes my duty to require your peaceable departure." Saint-Pierre took his time replying, eventually saying that he'd forward the message to Montreal.

Having completed his mission, Washington set out to return home on December 16. His journey through the wilderness was beset by difficulties and dangers. He almost drowned in French Creek; the Allegheny River was swollen and full of floating ice. There was deep, cold snow. He finally got back to Williamsburg on January 16, having traveled more than eight hundred miles in two and a half months.

Governor Dinwiddie then called up a militia regiment and sent Major Washington back to the disputed land, leading to the start of the French and Indian War. French forces were withdrawn from Fort Duquesne, Fort Venango and Fort Le Boeuf during the war and ordered north to lift a siege at Fort Niagara. French forces were defeated as they approached Fort Niagara. France lost the war, and England prevailed.

English troops were then stationed in Fort Venango, but they were massacred by Seneca Indians in 1763. The Senecas captured the fort during Pontiac's Rebellion. Not much happened in the region during the American Revolution, but a company of United States soldiers arrived in 1787. They built a new fort, Fort Franklin. It was a substantial wooden building, one and a half stories high and surrounded by a pine log picket. American troops were stationed at Fort Franklin until 1803.

It took several years for New York, Connecticut and Massachusetts to work out conflicting claims to the territory. Eventually, the Holland Land Company ended up owning all of Western New York and most of Western Pennsylvania. Joseph Ellicott served as its agent in Batavia, New York; Harm Jan Huidekoper got the job in Meadville, Pennsylvania. Boundary lines between New York and Pennsylvania were finalized after Pennsylvania bought the Erie Triangle in 1792.

New England settlers started coming down the Allegheny in the early 1800s. Others came over the hills from eastern and central Pennsylvania. A settlement grew up around Fort Franklin. The first steamboat, the *Duncan*, a stern-wheeler, chugged up the Allegheny River from Pittsburgh in 1826. Franklin became the head of navigation on the river.

Newcomers discovered iron ore in the hills and in bogs along the streams and rivers. Several took up the iron business. Seventeen smelting furnaces were operating by 1847. They produced about twelve thousand tons of "pig iron" each year, valued at the time at about $380,000. Water powered the furnaces, which burned charcoal.

Several settlers commented on a peculiar black stuff that oozed from the ground, coating the surface of springs. The stuff was found along Oil Creek, near Franklin, and at other places. Indians collected it, using it for medicine. Nobody knew what to make of it.

A few housewives gathered small amounts. They used it to treat cuts, bruises and burns. Some tried drinking it as a tonic. Eventually, a few druggists heard about it. They purchased samples, bottled it in small vials and sold it as "American Oil" or "Rock Oil." Nathaniel Carey, from Franklin, took more of an interest than most; he started gathering "Seneca Oil," putting it in five-gallon kegs and shipping it to Pittsburg.

Meadville, Crawford County's largest city, was settled in 1788. Titusville, the county's other city, was established on Oil Creek in 1796. It was settled by Jonathan Titus, a land agent from Blair County, Pennsylvania, and Samuel Kerr, a Holland Land Company surveyor. These gentlemen purchased 1,700 acres of land along the creek. Others joined them, creating a small settlement. Titus wanted to call his community Edinburg, but others insisted it be called Titusville. It was incorporated in 1847 as Titusville, Pennsylvania.

Lumbering and subsistence farming were the primary occupations when Ebenezer Brewer arrived. Moving his lumber business from Barnet, Vermont, he purchased several thousand acres of timberland along Oil Creek in the early 1840s. Once established, Ebenezer formed a new lumber company with local investors, creating Brewer, Watson & Company.

Titusville, Pennsylvania, is about sixty miles south of Fredonia, New York.

ROCK OIL: EARLY REPORTS

Sumerians used bitumen or crude oil about five thousand years ago. Persian military forces used oil-soaked flaming arrows during the siege of Athens in 480 BC. Herodotus described crude oil about forty years later. However, interest dropped off; there don't seem to be records of anyone doing much with it for the next few thousand years. The French missionary Père Joseph de La Roche found oil on the North American continent on July 18, 1627, when he came upon the Seneca Oil Spring near Cuba, New York. He called it *la fontaine de bitumen*.

A Jesuit missionary described the Indians' use of oil in 1656: "It is said that the oil is used by the Indians to anoint themselves and to grease their heads and bodies."

Captain Chabert Joncaire described the Cuba Oil Spring in his 1721 letter: "At a place the Iroquois called 'Ganos' there was a spring, the waters of which were like oil, and there taste like iron; and at a little distance from it there was another of the same character, the waters of which were used by the savages to cure all manner of disease."

Otherwise, the Cuba Oil Spring seems to have been somewhat ignored until Sir William Johnson, writing at Fort Niagara on September 19, 1767, reported, "Asenshan came in with a quantity of curious oyle, taken off the top of the water of some very small laeke near the village he belongs to."

Colonel Brodhead's division of General Sullivan's army visited the Seneca Oil Spring in 1779 on its way home from the Seneca campaign. He described the oil spring in his report and correspondence. Orasmus Turner is believed to have visited the Oil Spring sometime in the 1840s. He provided a description of the curiosity in his 1849 *History of the Holland Purchase*:

> *The celebrated Oil spring is two miles from the village of Cuba, on Oil Creek. Most readers are familiar with its peculiar character. It is a curious fact; and demonstrates how wide the range of the French Jesuits and traders was, over the region of Western New York; that Joncaire knew of the existence of this spring and described it to Charlevoix, in 1721. The mile square of land embracing it, was one of the reservations of the Seneca Indians, in their treaty with Robert Morris. The Indians regarded it of great value; attributed important medicinal qualities to the oil; in early years, after settlement commenced, it was a place with them, of frequent resort. They used to spread their blankets upon the water, wring them, collecting the oil in their brass kettles.*
>
> *Soon after the settlement of the country, the oil was collected and sold; and has been is use more or less, for nearly fifty years, though it is not certain that it possesses much virtue. The waters of the spring are pure and cold, not tainted with the oil. When the oil is skimmed off it will accumulate again, over the surface of the water, in one hour. It has a strong bituminous smell, in appearance, not unlike the British oil.*

The first recorded oil strike in Western New York occurred in 1832; it took place near Yorkshire corners, about twenty-five miles northwest of the Cuba Oil Spring. Chauncey Spear and Wells Cheney were trying to dig for coal. They dug three deep pits. Striking rock at about thirty feet in all three pits, they got a rush of black oil and water. Not finding coal and unable to dig deeper, they gave up the project. James Hall, the state geologist, wrote up an official description of their effort and published it in an appendix to the *New York State Geological Report* in 1840.

Oil was also found in other places. Nathaniel Carey, a tailor from Mansfield, Connecticut, discovered oil springs in the French Creek Valley. He took an early interest in the stuff, figuring that he might be able to sell some. Getting his brother to help, they gathered several kegs of oil. Taking their oil to the Pittsburgh market, they sold it as "Seneca Oil," claiming that it was a miraculous cure for a multitude of ailments.

A little later, General Hayes of Franklin, Pennsylvania, bought three barrels of oil. He shipped them to Baltimore. The merchants who received the product did not fancy the odor of the stuff or appearance of the barrels. They dumped the oil in Chesapeake Bay and burned the barrels.

Then Samuel Kier took an interest in oil.

Samuel Kier: Early Oil Enterprise

Samuel Martin Kier was the son of Thomas and Mary Martin Kier. He was born in Conemaugh Township in Indiana County, Pennsylvania, in 1813. The Kiers were Scotch-Irish immigrants who owned several salt wells around Livermore and nearby Saltsburg. Salt was an important commodity at the time. It was used to preserve meat and make dull or partially spoiled food somewhat edible.

Early settlers brought salt over the mountains from Philadelphia. Later on, merchants started bringing it in and selling it for about eight dollars per bushel. In the late 1790s, General James O'Hara discovered that salt could be purchased from the Onondaga Salt Works near Syracuse, New York, and brought to Western Pennsylvania. It cost about four dollars per bushel. However, it had to be shipped across Lake Ontario to the Niagara River. Then it had to be carted around Niagara Falls, shipped over Lake Erie to Presque Isle, carried over the Venango Trace and floated down French Creek and the Allegheny River to Pittsburgh. Wilkinson ran a salt boat on Lake Erie for a time, but the War of 1812 came along and interrupted the salt supply.

Another source had to be found. A small salt well or spring was discovered on the banks of the Conemaugh River near Saltsburg. William Johnston started operating a saltworks. He got salt by evaporating the water, but he needed a more productive supply. He tried drilling with a spring pole device. After drilling through two hundred feet of rock, Johnston struck a vein of salt water in 1813. Yanking his bit from the hole, he got a fountain of salt water spouting eight feet high in the air. Over the course of the next few years, other salt wells were drilled along the lower Allegheny.

The Kiers took up the salt business and expanded into other activities. Samuel helped found Kier, Royer & Company in 1838. This was a canalboat business shipping coal over Pennsylvania's Main Line, a system of canals, locks and railroads connecting Pittsburgh with Philadelphia. Samuel Kier then invested in coal mines, brickyards and a pottery factory.

Kier, along with other investors, including Benjamin F. Jones, founded a few iron factories in west-central Pennsylvania. This business eventually grew to become the forerunner of Jones and Laughlin Steel, one of the country's largest steel producers. Even so, salt and salt production continued to be Samuel's primary enterprise.

A dark, oily liquid started fouling some of Kier's salt wells in the 1840s. They called it "rock oil." At first, Kier simply skimmed it off the wells and dumped it in the nearby Main Line Canal. Canalboat operators complained about the stuff greasing their towlines and soiling their boats, but the salt manufacturers continued dumping the bothersome nuisance into the canal.

One afternoon, a few boys threw a burning branch in the canal, and the canal caught fire. The sight of the burning canal suggested that perhaps the stuff could be used for something—creating fire or producing light. Without formal training in science or chemistry, Samuel Kier decided to pursue the possibility. He started working with Professor James Curtis Booth to see if they could do something with the stuff, perhaps find or create some kind of product.

Professor Booth was a graduate of the University of Pennsylvania. He had also spent a year studying with Dr. Amos Eaton at the Rensselaer Polytechnic Institute. Professor Booth and Samuel Tier experimented with distillates of crude oil until they got a clear fluid. They called their product "Rock Oil," later changing the name to "Seneca Oil." Samuel Kier started bottling the fluid to be sold as a patent medicine. He charged fifty cents per bottle.

As part of his marketing strategy, Samuel hired several men to drive around the country in gaily decorated medicine wagons. They proclaimed the marvels of Kier's "Seneca Oil" while selling it to the public. They advertised their product:

> *Kier's Petroleum, or Rock Oil, is celebrated for its wonderful curative powers. A natural remedy! It's procured from a well in Allegheny County, Pa., four hundred feet below the earth's surface. Put up and sold by Samuel M. Kier, 363 Liberty Street, Pa.*

The healthful balm, from nature's secret spring,
The bloom of health and life to man will bring;
As from her depths the magic fluid flows,
To calm our sufferings and assuage our woes.

Samuel Kier continued experimenting with oil until he found a way to make kerosene. Kerosene had been known for some time, but it wasn't widely produced as it was considered to have little commercial value. Whale oil was the fuel of choice, but whale oil was becoming scarce and expensive.

Kier started selling kerosene, which he called "Carbon Oil," to local miners in 1851. He even developed and produced a lamp designed to burn his product. Kier didn't bother to patent his lamp or kerosene process. Even so, Kier's income grew to more than $40,000 per year, a huge sum at the time.

Samuel Kier set up an oil refinery, the world's first, in Pittsburgh on Seventh Street near Grant Street in 1853. Using a five-barrel still, Colonel Drake started producing illuminating oil from petroleum in 1854. He continued collecting crude oil by skimming his salt wells until he received his first shipment from Drake's well on Brewer property near Titusville in 1859.

EBENEZER BREWER AND VENANGO COUNTY: DIGGING FOR OIL

Ebenezer Brewer was born in 1789 and grew up in Henniker, New Hampshire. After serving in the militia during the War of 1812, he settled in Keene, New Hampshire. He established a glassworks and married Julia Emerson in 1817. Leaving New Hampshire, the young couple moved their family to McIndoe's Falls, Vermont. Ebenezer started out by keeping a public house, and then he got into the potash and lumber business.

Once in the business, he moved up to become the senior partner of Brewer, Gilchrist & Company. They became "wood choppers." They purchased land and cleared it, making potash and sawing logs into lumber. They floated their lumber down the Connecticut River and sold it in Hartford. Ebenezer acquired considerable wealth by doing this and became president of the Wells River Bank. After clearing the available land in Vermont, Ebenezer was forced to take his business elsewhere. Relocating to Pennsylvania, he purchased timberland along Oil Creek in Venango County.

The region had advantages. There was timber and possible mill sites. Oil Creek flowed into the Allegheny River. Lumber could be sawed on site and floated down the Allegheny to be sold at markets in Pittsburgh and elsewhere. Ebenezer packed up and moved to Titusville sometime during the 1840s. On site, he formed a partnership with local investors, including James Ryan, Jonathan Watson, Rexford Pierce and Elijah Newberry. They created Brewer, Watson & Company.

While building a sawmill, Ebenezer noted a peculiar phenomenon. There were oil springs on the property. These were an interesting curiosity. Then he noticed workmen skimming oil off the surface of the springs and using it to lubricate sawmill machinery. Others gathered the stuff and burned it, using it to create a smoky source of illumination. Some claimed that Seneca Indians drank it as a tonic.

Wishing to know more about this strange substance, Ebenezer had an employee gather up about five gallons. Ebenezer sent these back to his son, who happened to be practicing medicine in Vermont. Dr. Francis Brewer tried using some of it as a medicine, dispensing it to a few patients. Wanting to know more about it, Dr. Brewer sent a sample to Dr. Dixi Crosby, his chemistry professor at Dartmouth College.

Dr. Crosby took a look at the sample of "rock oil." Examining it, he figured that it might be useful if you could find enough to make it worthwhile. Dr. Brewer followed up by taking a trip to Titusville, where he contracted with J.D. Angier to collect oil. Angier dug a number of trenches to convey oil and water to a central basin, where he skimmed off the oil. He was able to collect three or four gallons per day by doing so, hardly enough to support a commercial venture.

Ebenezer became alarmed when his son got involved with a group of people who formed something called the Pennsylvania Rock Oil Company. The company paid $5,000 to sign a lease to "dig" for oil on a one-hundred-acre parcel of farmland along oil creek. Then the company spent more money purchasing rights to another twelve thousand acres. "You're associated with a set of sharpers." Ebenezer warned his son. "If they haven't already ruined you, they will surely do so if you are foolish enough to let them do it."

Ebenezer didn't believe that it was possible to collect or obtain enough "rock oil" to ever turn it into a paying proposition. Moreover, he had misgivings about the two New York lawyers who happened to be involved in the scheme: George Bissell and Jonathan Eveleth.

However, the company did turn a profit. Ebenezer sold his holdings in about 1860. Pocketing his profit, he moved down the Allegheny River to Allegheny City, now part of Pittsburgh. Going back into the lumber business, he opened a lumberyard and became a promoter of Pittsburgh. He took an active role in raising money to defend the city against a threatened Confederate attack in 1864, and Pittsburgh named a defensive fort after him: Fort Brewer. Ebenezer took an active interest in planning and building Pittsburgh's St. Andrews Episcopal Church, contributing $20,000.

DR. FRANCIS BREWER:
RESIDENT OF WESTFIELD, NEW YORK

Francis Beattie Brewer was born in Keene, New Hampshire, on October 20, 1820. His father, Ebenezer Brewer, left New Hampshire in 1822 and moved to Vermont. He became a "wood chopper," buying land, clearing forests and then moving on. After Ebenezer cleared his property in Vermont, he had to move. As there was available timberland in northwestern Pennsylvania, with an emerging lumber market in Pittsburgh, he moved to Pennsylvania.

Young Francis stayed behind. He graduated from Dartmouth College in 1843 and enrolled in the medical department at Dartmouth. He completed his medical education at Jefferson Medical School in Philadelphia. Then Francis went home, back to the Connecticut River Valley, to practice medicine. Dr. Brewer was practicing medicine in Barnet, Vermont, when he started receiving letters from his father. The letters described a peculiar substance encountered on the surface of some springs on his father's property.

Some called it creek oil, and others called it rock oil. Many believed it to be a remedy of great efficacy; they said it could be used to treat several diseases, including rheumatism, neuralgia and affections of the throat. Dr. Brewer asked his father to send a sample. Ebenezer had his workmen gather up several gallons of oil and then sent them on to his son.

This initiated a series of unforeseen events that culminated in unanticipated outcomes. Dr. Brewer discussed these events in an article that was printed in the *Titusville Morning Herald* on January 28, 1881:

> *Statement of Dr. Francis Brewer: I was practicing medicine among the hills of Northern Vermont in 1849 when I received from my late father, Ebenezer Brewer, who was then residing in Titusville, five gallons of "creek oil" as it was then called. This was before the days of the railroads, and*

This oil portrait of Dr. Francis Brewer hangs in the Chautauqua County Historical Center Museum housed in the McClurg Mansion in Westfield, New York. *Courtesy of Chautauqua County Historical Society. Photograph by Niles Dening.*

it was carried by Jon Lock, by stage, by canal, and by private conveyance, until delivered to me, with the assurance it possessed great medicinal and curative properties.

I at once gave it a trial, and was not disappointed in its healing virtues, and so long as I continued in my profession, I had it in constant use. I sent some of it to Dr. Dixi Crosby, then at the head of the New Hampshire Medical Schools. I also gave some to Dr. Hubbard, Professor of Chemistry in Dartmouth College, who, on a hasty examination, pronounced it a very valuable oil, produced at a great depth in the earth, by the decomposition of vegetable and bituminous matter, thrown up in the form of a gas, condensed upon the overlying rocks and forced through the fissures as rock oil, remarking it, would probably never be found in sufficient quantities to entitle it to a commercial value.

Albert Crosby, Esq., son of Dr. Crosby, took some interest in the specimen and exhibited it to several scientific gentlemen, among others, George H. Bissell, Esq., and Jonathan Eveleth, Esq., both educated men and lawyers in New York City. Three years later I moved to Titusville and became a member of the firm of Brewer, Watson & Company and for the first time examined the oil spring, in the vicinity of one of our saw mills, just below the village, and called the "upper mill." I became satisfied there was oil in abundance.

It was often a subject of discussion between us: Mr. Jonathan Watson, Mr. J.D. Angier, and I about how we could use this oil and make it profitable. Our firm, Brewer, Watson, and Company leased the spring to Mr. Angier on July 4, 1853. He wished to work the spring more efficiently, perhaps make further excavations, and utilize the adjoining territory, believing as he did at the time, the oil to be a product of coal in the higher ground along the stream. This was probably the first oil lease ever made.

In pursuance of this arrangement, ditches were opened and trenches were dug. The oil and water were conveyed to a central point, where by some simple device the separation took place, and the oil supply materially increased. We used it for lighting the mills and lubricating the machinery.

I visited New England in 1844 where I met Mr. Crosby, of whom I have spoken, and invited him to come to Pennsylvania to assist in organizing a company to work these wells and introduce our new and valuable product to a world market. He came, and from the first was enthusiastic in his estimate of its value and abundance. I visited several localities along the stream with him where oil appeared, as far down as to the McClintock farm. Here was a well in the middle of the creek. Crosby got excited and recommended the purchase of the entire oil district then known. I suppose that $50,000 invested at that time would have ended up controlling the entire oil basin on both sides of the stream.

As we stood on a circle of rough logs surrounding the spring and saw oil bubbling up, spreading its bright and golden colors over the surface of the water, it seemed like a golden vision. Crosby at once proposed to *purchase the whole farm, which we could have done for $7,000. But as our pecuniary ability was limited to a much smaller sum, I was obliged to decline the tempting opportunity. When I told him we did not wish to take capital from our lumber business to take a chance on oil, he said, "Damn lumber, I would rather have McClintock's farm than all the timber in western Pennsylvania."*

However, he obtained from McClintock an option to the land on the east side of Oil Creek for $1,000, on the west for $5,000, and the spring for $2,000, or for all of it, $7,000. This, probably, was the first "option" for the purchase or sale of oil property, as such, ever made. In a few years, it became common. It was a verbal agreement—buyer's option—thirty days between Crosby, of New Hampshire, and McClintock, of Pennsylvania, made in August 1854. I was a witness.

On our return to Titusville an agreement was entered into with Crosby by myself, representing the firm of Brewer, Watson, & Company, in which it was stipulated that if Crosby would find parties to take hold of the business and furnish capital to work the territory, we would sell the Hibbard farm, containing 100 acres of land and on which the original well was being worked, and on which the first, or Drake well, was bored and completed six years later.

The conditions of the contract were $5,000 for this land, and the oil found in our other lands, consisting of several thousand acres. A stock company was to be organized with a capital of $250,000, of which one-fifth was to belong to Brewer, Watson, & Company, one-fifth to be treasury stock, and three-fifths to be placed where it would do the most good. With several specimens of oil, Mr. Crosby left Titusville full of hope, saw two gentlemen of whom I have spoken—Mr. J.G. Eveleth and George H. Bissell—exhibited the samples of oil and submitted the contract, which he wrote me, met with their hearty approval.

They cordially consented to the conditions; and on their return to New York, which would be made in a few weeks, they would make every arrangements necessary to carry out their part of the contract. Meantime, I was asked to send on some oil and meet them in the city at an early date, to be named by them.

Three barrels of crude petroleum were gathered on the lands owned by Brewer, Watson & Company. J.D. Angier collected this oil from a rude excavation curbed with rough, unhewn logs. The barrels were carried to Erie, Pennsylvania, by J.J. Hurst. From Erie, they went to Buffalo by steamer and then by canal to New York City. They were consigned to the firm of Eveleth & Bissell at Broadway and Franklin.

Several letters were exchanged. Dr. Brewer wrote a letter to Eveleth & Bissell in the summer of 1854, describing the substance as a "peculiar oil, surpassing in value any other oil now in use for burning, for lubricating machinery, and as a medicinal agent. The yield is abundant and the supply inexhaustible." The letter to Eveleth & Bissell also contained a proposition from Brewer, Watson & Company. Dr. Brewer suggested creating a New York company to produce and market oil.

Soon afterward, Dr. Brewer traveled to New York. He took another sample of oil with him. The sample was submitted to experts for analysis. They agreed on the quality of the oil, but no one believed that sufficient quantities could be found to market it as a product. Dr. Brewer believed that it could. Eveleth & Bissell considered the matter, and then they wrote Dr. Brewer a letter on November 6, 1854, telling him that they were organizing a joint stock company.

The Pennsylvania Rock Oil Company was organized as a New York State corporation later that month. Needing an assessment of the value of the product, Eveleth & Bissell sent a considerable quantity of oil to Professor Benjamin Silliman at Yale University. Professor Silliman was a Yale graduate holding a bachelor's and master's degree. He had studied chemistry with Professor Woodhouse at the University of Pennsylvania before going back to Yale as the college's first professor of chemistry.

Professor Silliman was probably the most prominent chemist in the country at the time. The professor gave the oil sample a thorough analysis and published his findings in a scientific paper. His paper caught the attention of several investors, but few invested. As the enterprise was chartered in New York, this established an element of stockholder liability. Moreover, there were concerns about the practicality, reliability and feasibility of such an enterprise.

James Townsend, president of the New Haven City Bank, suggested restructuring the company under Connecticut law in September 1855. The new company was chartered as the Penn Rock Oil Company of New Haven, Connecticut. Professor Silliman helped establish credibility by becoming president. Colonel Edwin Drake and others joined the enterprise. The Penn Rock Oil Company leased Brewer land near Titusville in 1856.

Colonel Drake struck oil on August 30, 1859. Dr. Brewer then spent most of his time and effort on the oil business. As Brewer, Watson & Company owned several tracts of land along Oil Creek, it leased the property to various firms and individuals, generating substantial profit. Dr. Brewer left

Left: This street sign marks Brewer Place in Westfield, New York. Brewer Place is a short street named in honor of Dr. Francis Brewer. He owned the land, as it was part of his estate. *Photograph by Hoffman.*

Below: This is a contemporary photo of Dr. Brewer's house in Westfield, New York. *Photograph by Hoffman.*

Pennsylvania on May 1, 1861, and moved to Westfield, New York. However, he continued his interest in the oil business until 1864, at which time the firm sold its Pennsylvania property.

Dr. Brewer took an active interest in Westfield. He purchased several tracts of land in and around the village. He organized the First National Bank of Westfield and started the Townsend Manufacturing Company, a large manufacturing business. Dr. Brewer became sole owner and president of the company in 1870. He then changed the name of the company to Westfield Lock Works. He served in the New York State legislature for two years (1873 and 1874). Dr. Brewer was elected to the United States Congress in 1882 and served one term, representing Western New York.

Dr. Brewer's lovely old home still stands on Brewer Place, a street in Westfield, New York. The memory still persists of Dr. Brewer as a kindly and Christian pillar of the Presbyterian Church. A special section of the *New York Times* on May 31, 1959, and the *Centennial Issue of the Petroleum Institute Quarterly* gave Dr. Brewer credit for being an important link in the development of today's petroleum industry: energy and power.

A contemporary (J.T. Henry) described Dr. Brewer as a "gentleman of solid worth, and we may add, of solid wealth, as well. He is eminently a man of the people; universally respected for his integrity and purity of character."

George Bissell: A Fortunate Occurrence

George H. Bissell was born at Hanover, New Hampshire. His father died when he was twelve. Left to his own resources, he managed to graduate from Dartmouth College in 1845. He took a job teaching Greek and Latin at the University at Norwich. Giving up this position, he traveled to Washington, D.C., where he got a job as a newspaper correspondent for the *Richmond Whig*. Then he traveled to Cuba in the spring of 1846 and then on to New Orleans, where he became connected with the *New Orleans Delta*'s editorial department.

He left that job to become the principal of the city's new high school. Then he moved up, becoming superintendent of the New Orleans Public Schools. He served in that capacity while studying law. He continued working as superintendent until poor health necessitated his return to the cooler climate of New England. While visiting Professor Crosby at Dartmouth College, he happened to notice a bottle of petroleum.

George H. Bissell's account of this event, what happened and how it led to the oil industry can be found in the *Report of the United States Revenue Commission on Petroleum as a Source of National Revenue.* The February 1866 report is found in House Executive Document N. 51, thirty-ninth Congress, first session:

George Bissell's account—In the year 1853, I saw at the office of Professor Crosby, of Dartmouth College, a bottle of petroleum given him by Dr. Brewer, of Titusville, Pennsylvania. It was found on Dr. Brewer's land on Oil Creek. I became greatly interested in the product, and, about six months after, went to Titusville, with Mr. J.C. Eveleth, who was then, and had been previously, my partner in other business.

We bought together from Brewer, Watson, & Company, what were then thought to be the principal oil lands in Pennsylvania. They were in extent one hundred acres in fee simple, and one hundred and twelve acres on lease for ninety-nine years, on Oil Creek, about two and a half miles below Titusville, for which lands we paid $5,000. Before purchasing, we prospected the land. We dug holes in the ground, six or seven feet deep. The oil and water together percolated into these holes, and the oil was afterwards gathered by dipping woolen cloths into the mixture, and wringing the cloths out. In three or four hours, one of the holes would collect from a pint to a quart of oil.

We did not prospect the oil for medicinal purposes, but we believed it might make a good illuminator, and we sought it as an article of commerce. Illuminating oil from coal was just beginning to be talked of, but very little was made then. We then, in 1854, organized a company in New York City under the name of the Pennsylvania Rock Oil Company. The nominal capital was $500,000. This was the first petroleum company ever organized in the United States, or elsewhere, so far as I know.

We proceeded to develop these lands by trenching them and raising the surface oil and water into vats. These trenches varied from twelve to eighteen feet deep, three to four feet wide, and almost sixty or seventy feet in length. They were dug to converge, increasing in depth, to a central point at a small saw-mill upon our lands. We collected the oil by a pump worked by the water power connected with the mill.

The supply of oil was very limited, amounting to, perhaps, a few barrels in the course of a season, which we sold to parties for $1.50 a gallon. They retailed it for medicinal purposes. We conducted these operations for three years before employing Dr. Silliman, of Yale College, to analyze the oil. We furnished him with all the useful apparatus required for his experiments.

He was engaged for about four months and published his report in the fall of 1855. The report excited attention in New Haven and some gentlemen in that city proposed to take an interest in our company on condition that the company be reorganized in New Haven. This was done and Professor Silliman was elected president.

The work on trenching the lands continued until we heard that Mr. Kier of Pittsburg had obtained a small quantity of oil from a salt-well. After seeing pictures of derrick drilling for salt, I came up with the idea of perhaps drilling for oil rather than trying to mine it. Others weren't so enthusiastic.

However, George Bissell managed to convince the others to try sinking a well. They needed to find a person who would go to Titusville and do it. They decided on Edwin Drake. He was a stockholder and former railroad conductor. Drake still held a railroad pass that entitled him to free travel on any and all railroads. However, James Townsend insisted that Edward Drake needed some sort of title to impress the local people in Pennsylvania. "Colonel" sounded good, so Edwin Drake, an unemployed railroad conductor, became Colonel Drake. Then they sent him to Titusville.

Using his railroad pass, Colonel Drake took the New York and Erie Railroad to Dunkirk, New York. Then he rode the Buffalo and Erie Railroad to Erie, Pennsylvania. He took a coach to get to Titusville. When he got to Titusville, he surveyed the scene and hired one driller and then another. Even though he was delayed by a series of obstacles, Colonel Drake managed to get oil at sixty-nine and a half feet on Sunday, August 28, 1859.

JONATHAN GREENLEAF EVELETH:
ATTORNEY AND BUSINESSMAN

Jonathan Greenleaf Eveleth was born in New Gloucester, Maine, in 1821. His father, James Eveleth, was a leather tanner and boot maker. Jonathan was a direct descendant of several old Massachusetts colonial families, including Joseph Eveleth, a juror in the Salem witch trials, and Theophilus Parsons, chief justice of the Massachusetts Supreme Judicial Court.

Jonathan graduated from Bowdoin College in 1847 and Harvard Law School in 1854. After completing law school, he worked at the Fessenden & Deblois Law Firm in Portland for a time before moving to New York City.

When he got to New York, he formed a partnership with George Bissell. They opened an office at 14 Wall Street but closed it in 1859 to take up the oil business.

Bissell and Eveleth invested $5,000 to purchase and lease more than two hundred acres of land in what they believed would be the principal oil fields on and around Oil Creek. The company started out by trying to collect oil in trenches; it sold small quantities of this oil at high prices ($1.50 per gallon) for medicinal purposes.

Wishing to generate more interest with possible investors, Eveleth and Bissell hired Professor Benjamin Silliman, the Yale chemist, to investigate and analyze their product. As Professor Silliman's primary interest was the emerging science of fractional distillation, he used this to conduct his analysis.

Silliman submitted his historic report in 1855. The report concluded, "Gentlemen…there is much ground for encouragement in the belief that your company has in their possession a raw material from which they may manufacture…very valuable products." The report got the attention of James M. Townsend, president of the New Haven City Savings Bank, as well as other investors.

James Townsend urged Eveleth and Bissell to give up their New York company and reincorporate the company in Connecticut. Under New York law, a stockholder's property could be seized to cover company debts. This wasn't the case in Connecticut—stockholders were not liable for debts incurred by a company. The Pennsylvania Rock Oil Company of Connecticut was therefore reincorporated in that state on September 18, 1855. Capital was subscribed in a short time by investors in and around New Haven.

James Townsend convinced Edwin Drake, a New York and New Haven Railroad conductor, to buy some stock in the company. When Drake became ill during the summer of 1857, he gave up his railroad job. Even though Drake wasn't working, he still had a railroad pass. Townsend got him to use his pass to take a trip to Titusville to check out the property. Edwin Drake, as Colonel Drake, went to Titusville. He examined the property, talked with some people, came back and wrote a report of his findings.

On the basis of Drake's report, James Townsend made the decision to create a new company. The Seneca Oil Company was chartered on March 23, 1858. Being majority stockholders of their old Pennsylvania Rock Oil Company, Townsend, Eveleth and Bissell leased the old company's properties to themselves, and they became the principal stockholders of the new company. Professor Silliman added a bit of credibility by becoming president.

Even so, Eveleth and Bissell maintained controlling interest. They picked Edwin Drake to be their general agent. The new Seneca Oil Company gave him an annual salary of $1,000 and sent him back to Titusville, this time to get oil.

Eveleth and Bissell continued buying up land along and around Oil Creek, but Eveleth's life was cut short by illness. He died in New York City on December 10, 1861. Jonathan Eveleth was later described by some of his contemporaries as "sage, cautious, and tenacious" and the "principal promoter" of the company. He made many important decisions. Some have called him the "father of the American oil industry."

EXCERPTS FROM PROFESSOR SILLIMAN'S REPORT: CHEMICAL ANALYSIS AND OBSERVATION

Professor Silliman's report was issued on April 16, 1855. He concluded:

I have submitted this lamp burning petroleum to the inspection of the most experienced lampists who were accessible to me, and their testimony was, that the lamp burning this fluid gave as much light as any which they have seen, that the oil spent more economically, and the uniformity of the light was greater than in camphene, burning for twelve hours without a sensible diminution, and without smoke. I was, however, anxious to test the amount of light given, more accurately than could be done by a comparison of opinions.

With your approbation, I proceeded therefore to have constructed a photometer, or apparatus for the measurement of light, upon an improved plan. Messrs. Grunow, scientific artists of this city, undertook to construct this apparatus, and have done so to my entire satisfaction. This apparatus I shall describe elsewhere—its results only are interesting here. By its means I have brought the petroleum light into rigid comparison with the most important means of artificial illumination. Let us briefly recapitulate the results of these…

The unit adopted for comparison of illumination is Judd's Patent Sixes Sperm Candle. The sperm oil used was from Edward Mott Robinson, of New Bedford—the best winter sperm remaining fluid at 32 degree Fahrenheit. The colza oil and Carcel's lamps were furnished by

Dardonville, lampist, Broadway, New York. The gas used was that of the New Haven Gas Light Company, made from best Newcastle coal, and of fair average quality.

The distance between the standard candle, and the illuminator sought to be determined, was constantly 150 inches—the photometer traversed the graduated bar in such a manner as to read, at any point where equality of illumination was produced, the ratio between the two lights. I quote single examples of the average results.

Rock-oil was somewhat superior in illuminating power to Carcel's lamp of the same size, burning the most costly of oils. It was also equal to the "Diamond Light" from a lamp of one-half greater power, and consequently is superior to it in the same ratio in lamps of equal power. The camphene lamp appears to be about one-fifth superior to it, but, on the other hand, the rock-oil surpasses the camphene by more than one-half in economy of consumption.

Compared with sylvic oil and sperm oil, the rock-oil gave the whitest disc of illumination. Compared with gas, rock-oil gave more light than any burner used except the costly Argand consuming ten feet of gas per hour. To compare the cost of these several fluids with each other, we know the price of the several articles, and this varies very much in different places. Thus gas in New Haven costs $4 per 1,000 feet, and in New York $3.50 per 1,000, in Philadelphia $2.00 per 1,000, and in Boston about the same.

Such sperm oil as was used costs $2.50 per gallon, the colza about $2, the sylvic oil fifty cents, and the camphene sixty-eight cents. No price has been fixed upon for rectified rock-oil. I cannot refrain from expressing my satisfaction at the results of these photometric experiments, since they have given the oil of your company a much higher value as an illuminator than I had dared to hope. In conclusion, gentlemen, it appears to me that there is much ground for encouragement in the belief that your company has in their possession a raw material from which, by simple and not expensive process, they may manufacture very valuable products.

COLONEL EDWIN L. DRAKE: DRILLING FOR OIL

Edwin Laurentine Drake was born on March 29, 1819, in Greenville, Green County, New York. His family moved to Vermont when Edwin was eight years old, and he grew up in Castleton Corners. He worked a number of jobs in and around Castleton before getting a job as a conductor on the New York and New Haven Railroad. Moving to New Haven in 1854, Drake met James Townsend, a banker and founder of the Seneca Oil Company. Townsend convinced Drake to invest in the company.

Drake bought $200 worth of stock. The company purportedly held a lease on 105 acres of land near Titusville, Pennsylvania. Drake was told that there were oil springs on the property, but there were some concerns about the validity of the lease. Wanting to check the title, Townsend, Pierpont, Ives and Bowditch prevailed on Drake to use his railroad pass to go to Pennsylvania to check it out.

As they offered Drake a salary of $1,000 per year, Drake used his railroad pass to go to Titusville. He got there in December 1857. Heavy rains had fallen, and "winter was closing in with chill and mud, as it can only on Oil Creek." Drake's attention was first directed to oil while riding on a coach from Erie to Titusville. The driver spoke of a place called "Oil Creek."

Colonel Drake asked the driver, "Why is it called Oil Creek?"

"There's oil in it," was the reply.

When Drake got to Titusville, he discovered that oil was being skimmed from the creek and a few springs near the creek. Workmen were using it as a lubricator and burning it for illumination. He was shown a bottle of Seneca Oil in a Titusville store. Becoming interested in the substance, he visited the Brewer & Watson sawmill. Oil was being used to lubricate the journals of a large circular saw turning 1,200 revolutions per minute. Oil burned in torches to light the mill. Workmen told him that their rock oil was a better lubricant than expensive whale oil. As a lubricator alone, Colonel Drake figured that there was a fortune to be made with this oil.

Dr. Francis Brewer took Colonel Drake to the spring from which oil for the mill was taken. It was a rusty, disgusting-looking pool. Dr. Brewer spread a blanket over the surface of the pool and then gathered it up. The blanket dripped with dark fluid. Dr. Brewer wrung it out, getting a good half pint of oil from the blanket. Half a bucket or more of oil was thus secured.

Satisfied as to where "Seneca Oil" could be obtained and its value, the question became whether enough oil be gathered to market it as a viable product. Before Colonel Drake left Titusville, he concluded that sufficient quantities might be obtained by boring for it. Most regarded his scheme as a delusion of insanity. He was, however, undaunted. Drake went back to New Haven determined to go back and drill for oil. When he came back to Titusville, he came back as the agent for a new corporation, the Seneca Oil Company.

Colonel Drake was a genial, bearded gentleman of forty. He was a striking figure on the streets of Titusville with his long frock coat, loud-checkered trousers and high silk hat. More effective than his attire were his affability and natural charm, which together with his good-humored storytelling made him a popular gentleman about town.

However, he ran into an unforeseen problem when he got ready to start drilling: he wasn't able to hire anyone "to work for a lunatic." In some distress, Drake traveled to Tarentum, about one hundred miles away. He hired a salt borer, agreeing to give the salt borer at least one thousand feet of drilling to do. The contracted price was $1.25 per foot for the first hundred feet and then $1.00 per foot after reaching that depth.

The salt borer said that he'd be there in August, but he didn't show up. Colonel Drake went down again, and the fellow suggested that he might be able to get there by September. Again, he didn't show up. The now thoroughly vexed Colonel Drake traveled the hundred miles another time. While staying in the hotel at Tarentum, Colonel Drake accidentally found out why the fellow had disappointed him.

The salt borer had never meant to go to Titusville at all. Regarding Drake as crazy, he figured, "The easiest way to get rid of him was to make a contract and pretend he was going to come." After ransacking the country for several weeks, Colonel Drake found another salt borer who agreed to work for him. That man got as far as Pittsburgh but then died. It was now October, late in the season. Lewis B. Peterson, a salt manufacturer at Tarentum, became interested in Colonel Drake. He advised Drake to hold off until spring.

In April 1859, Mr. Peterson advised Colonel Drake that he had a man for him. Colonel Drake got in touch with William Smith of Butler County and hired him to drill a well. Billy Smith brought one of his sons with him, and the two of them worked for two dollars per day. Two more sons came on the job a few days later. Colonel Drake furnished the tools. As was the custom, effort was made to crib the well down to the rock. Water, however, gained too rapidly, and the cribbing had to be abandoned.

Colonel Drake took a trip to Erie and ordered ten-foot sections of cast-iron tubing from Liddle, Kepler & Company. He proposed driving sections of pipe into the rock. Drake ordered one-and-a-half-inch-thick pipe that was three inches in diameter. When the pipe was delivered on site, it was only half an inch thick and three inches in diameter. The manufacturer claimed that it couldn't be driven, and if driven, he couldn't bore through it. Everywhere he turned, Colonel Drake encountered discouragement.

Considerable time had been lost, and Drake decided to drive the half-inch pipe. The first section went home, but the second section broke the first section. He had to go back to Erie and purchase sections of pipe made from soft iron, one and a half inches thick. These were cast. The soft iron pipe drove all right, striking rock at thirty-one feet. Boring started on July 1, 1859. A depth of sixty-nine and half feet was reached on Saturday night, August 29, 1859. Work was suspended to mark the Sabbath, a day scrupulously observed by Colonel Drake.

Billy Smith walked out to the well on Sunday, August 30, 1859, and found it full of oil. He dipped up a barrel, and Colonel Drake's insanity dissipated. The Drake Well never flowed, but Billy rigged up a pump. The well surpassed expectation, producing twenty-five barrels of oil per day. Needing something to hold all this oil, they constructed vats, each capable of holding about six hundred barrels.

They needed barrels to transport the oil to get it to market. There were empty whiskey barrels in about every cellar in the region. These made excellent oil barrels. After Drake got his vats and about three hundred barrels filled, the drilling shack and well caught fire. The whole thing—vats, barrels, derrick and engine—burned to the ground. A gentleman on horseback rushed to tell Drake that his works had "burned up, slick and clean, engine and all."

"And the hole?" Colonel Drake asked. "Did it burn?"

"The fellow got mad," Colonel Drake reported, "but it was a relief to me. So long as the well remained, the works could be rebuilt." The works were rebuilt in thirty days, but they had to find a market for the oil. One hundred barrels were sold to a man named McKeown, a druggist in Pittsburgh, who refined oil as an experiment. Wanting to find more markets for all this oil, company representatives visited Erie, Cleveland, Cincinnati, Pittsburgh, Philadelphia and New York. This generated some interest.

Nevertheless, Colonel Drake's efforts were successful. His well yielded more than three thousand barrels of oil. He drilled another well but didn't get much oil. However, the citizens of Titusville elected him to become a justice of the peace. He lived in Titusville until 1863. Then he sold his Titusville house and moved to New York City, where he tried selling oil stock and land, but he had lost most of his money by 1866. His health was failing. A public meeting was held in Titusville on Wednesday evening, December 1, 1869, to come up with proper and appropriate measures for the relief of

The Drake Well replica is located on the grounds of the Drake Well Museum near Titusville, Pennsylvania. It is a "board for board" reconstruction of the original engine house as it appeared in the 1860s. The original building was destroyed by fire in October 1859. It was replaced one month later. *Photograph by Hoffman.*

Colonel Drake and his family. The residents of Titusville took up a collection, gave him the money and eventually prevailed on the Pennsylvania General Assembly to provide him with an annual $1,500 pension.

A South Bethlehem correspondent of the *Philadelphia Times* interviewed Colonel E.L. Drake in September 1879. A description of Colonel Drake was printed in the *Titusville Morning Herald* on September 11, 1879:

Above: Edwin Drake died in Bethlehem, Pennsylvania, in 1880. His body was exhumed and brought in 1902 to Titusville, where he was buried at Woodlawn Cemetery. Henry R. Rogers, a Standard Oil executive, erected this memorial to honor Drake. *Photograph by Hoffman.*

Left: The Drake Well Museum tells the story of the beginning of the modern oil industry with orientation videos, exhibits, operating oil field machinery and historic buildings in a park setting. Visitors enjoy a variety of special events, lectures and educational programs. *Photograph by Hoffman.*

Colonel Drake is afflicted with muscular neuralgia and has been an invalid since 1860. Though emaciated by the disease, there is still evidence of a noble frame and iron constitution—in fact none but the strongest could have sustained life in the many years of suffering that fell to the lot of this man. In stature Colonel Drake measures five feet nine and a half inches, and he is blessed with pleasant, honest, open, fearless features.

Edwin Drake died in Bethlehem, Pennsylvania, on November 9, 1880. His body was later moved to Titusville, where it remains. Henry Rogers, a Standard Oil executive, built a statue of Drake and memorial to mark the gravesite. The Drake Well Museum now stands near the site of Drake's first well. Drake's tools and artifacts are displayed in the museum.

DRILLING MORE WELLS: RUSH TO RICHES

News of Drake's success spread rapidly. Within twenty-four hours, hundreds of people were milling around the Drake Well. An eyewitness wrote that the excitement was fully equal to what he had seen in California at the time of the gold rush. Hundreds rushed to Oil Creek to lease land or buy it to drill a well.

A wave of excitement spread throughout the region, spilling over the state line into nearby Western New York. Chautauqua County was right next door, less than thirty miles from the oil discoveries. County residents were fascinated by the findings. Light from burning gas and oil wells in nearby Pennsylvania was nightly sighted in their sky. Excited by the very real possibility of grabbing sudden great wealth, several Chautauqua County residents responded to the call. Spilling over the state line, they rushed to get in on the frenzy.

Cyrus D. Angell from Forestville, New York, took charge of the Belle Island Petroleum Company. He came up with the idea that petroleum deposits were confined to belts. After drilling several wells around Franklin and making other observations, he came to that conclusion. He believed that two of these belts existed: one ran from Petroleum Centre to Scrubgrass and the other from St. Petersburg to Butler County. Most producers regarded Angell's theory with considerable skepticism but were influenced by his observations. These resulted in oil discoveries in nearby counties.

Haskell L. Taylor came from Stockton, New York. He created H.L. Taylor & Company, which, for a time, was the country's largest oil producer. His company owned more than three hundred wells by 1874, including "The Boss," on the Parker Farm near Criswell. This well produced more than two thousand barrels of oil per day. Preston Barmore, from Forestville, and Henry Rouse, from Westfield, rushed to get in on the action.

As the original Drake Well was located on Oil Creek, most believed that the best place to drill for oil was on the lowland, as close as possible to the creek. Therefore, there was a mad rush to secure the land around the Drake Well and all along the creek. Bissell bought all the stock of the Pennsylvania Rock Oil Company that he could buy and rushed to Titusville. He rushed around, leasing or buying farm after farm all along Oil Creek and the Allegheny River.

Available land bordering Oil Creek was soon taken up, and within a short time, the entire valley, even back to the hillsides, had been leased or purchased. Colonel Drake was advised to get in on the rush to purchase or lease land, but he ignored the advice. After several other wells were struck, Drake realized his mistake, but it was too late.

William Barnsdale and Henry Rouse from Westfield, New York, drilled the second well. Their well was located on the Parker farm, a short distance from the Drake Well. They got oil in November, but their yield was less than five barrels per day. They continued drilling. A few days later, when their well was down to 112 feet, oil rushed up, flowing over the top of the pipe. It produced about ten barrels of oil per day. They called it Barnsdall's Well, and it became a center of attention.

Brewer, Watson & Company started putting down a well on the McClintock farm at the lower end of Oil Creek. It hit oil in November 1859. David Crossley, from Titusville, started drilling on the opposite side of Oil Creek about one mile below the Drake Well. He got oil on March 4, 1860. The "B.W. & C. Well" pumped seventy-five to eighty barrels per day. The oil rush was on. People poured into Titusville.

A.B. Funk, from Warren County, struck oil in May 1861. His "Fountain Well," about seven miles below Titusville, startled observers by flowing three hundred barrels per day every day. Skeptics called it the "Oil Creek Humbug." They expected that the flow would soon cease, but it didn't. The well flowed for fifteen months. Mr. Funk cleared about $2.5 million on this venture. That generated more interest.

The "Empire Well," near the Funk Well, started flowing in September 1861. It flowed at the rate of three thousand barrels per day. Unable to find enough barrels to hold all the oil, the owners tried to block the flow. They couldn't. They threw up an enclosure around the well, but the oil kept coming. Refusing to be dammed, it overflowed the enclosure, flowing into Oil Creek. The creek was covered for miles. The yield bewildered the owners, glutting the market.

A few weeks later, the "Phillips Well" on the James Tarr farm started flowing, producing four thousand barrels of oil per day. The flow was so great that the owners built huge underground wooden tanks to contain the oil. Production boomed to five thousand barrels a day and then increased to more than six thousand barrels per day.

Production was monstrous, and it flooded the market. This could not be endured. Prices dropped, plummeting to ten cents per barrel. Coopers couldn't make barrels fast enough to keep up with the demand. Confronted with an economic crisis, a group of landowners and producers gathered at Rouseville, just north of Oil City, on November 14, 1861, to organize and take measures to increase the price of oil. They formed the Oil Creek Association to regulate production and establish a base price. The association limited production, stabilizing prices at four dollars per barrel.

The "Sherman Well" started flowing in 1862 and flowed until 1864. The "Noble and Delamater Well" started flowing in 1863. The "Craft Well," drilled a little later on the same farm, produced more than 100,000 barrels of oil. The most remarkable wells were the "Empire" and "Crocker." The "Crocker" started out by producing 2,500 barrels per day. It flowed for almost four months before dropping off to about 300 barrels per day.

With all of that oil coming out of the ground, someone had to find a way to get it to market. Oilmen weren't really that interested in that part of their business; they just wanted to find oil. After they found oil, they needed to put it in something to get it to market. They settled on forty-gallon wooden barrels, plus two gallons to make up for loss during transportation. Coopers from Pittsburgh and elsewhere rushed to the oil fields to make barrels.

Getting the barrels of crude oil to a refinery required wagons, teams and teamsters. Local farmers came to the rescue by hiring out their wagons. The delivery job frequently fell to farmers' older boys. These young men could usually get four or five of the forty-two-gallon barrels on a farm wagon. As they got about three dollars per barrel, many decided to give up farming. They became professional teamsters.

They hauled oil to the nearest railroad, where it was dumped in large, upright, open wooden cisterns mounted on flatcars. The Buffalo and Oil Creek Cross-Cut Railroad connected with the Dunkirk, Allegheny Valley and Pittsburgh Railroad. This railroad ran through Chautauqua County on its way to Joseph Dudley's Empire Oil Works in Buffalo, New York, and the large Atlas Refinery later on. The refinery and railroad were built to serve the oil industry.

Oil trains sometimes caught fire. If this occurred during the night, the country for miles around would be lit by the flames. This once happened on the Dunkirk, Alleghany Valley and Pittsburgh Railroad as a loaded oil train chugged through Wheelers Gulf, in the town of Pomfret, near Fredonia, New York.

Production on the original oil fields peaked in 1862, declining in 1863. Older wells started drying up, but new wells were coming on line. General Lee's invasion of Pennsylvania that June created such a ruckus that there was almost a total suspension of business. This decline in oil production created a spike in demand.

Oil prices soared. Oil reached $7.25 per barrel in September 1863. Buyers rushed to the region trying to buy oil, willing to pay the price.

Allegheny Valley Oil: Creating an Industry

There was a rush to Oil Creek to buy or lease land and drill a well. Each newly completed well in 1860—such as the Barnsdall, Crossley and Williams—increased the excitement, but they were nothing compared to the flowing wells, such as the Empire, Phillips, Woodford and Sherman. Production varied from 1,500 to 4,000 barrels per day. This was more oil than anyone had ever seen. It was bewildering.

When the wells along Oil Creek started producing large quantities of petroleum, strenuous efforts needed to be made to find a market for all of this oil. Continuous drilling flooded the market. At first, this overwhelming quantity of oil seemed to be of little value, but this changed as petroleum's fame as an illuminant spread throughout the world. People wanted light, and this generated a demand for kerosene.

The birth of the new industry—the petroleum industry—created refineries, the need for thousands of teamsters, railroad extension into the oil region and the spread of oil production up and down the Allegany Valley. T.S. Scoville, an early visitor to the region, described Oil City in 1861:

I found mud everywhere from four to six inches deep. Everything is muddy and dirty. Hotels crammed full, two in a bed everywhere, and three if they can get them in, not mention the number of small-stock travelers that pile in with the rest. Hotels not plastered. Buildings rough outside and in, set on stilts, all new, all hurried; great preparations for drilling, pumping, buying, selling, building—all excitement, life, and activity. At Tidioute there are some 200 wells in progress, and all the way from there here, thirty miles by raft, one is not out of sight of derricks and wells, hundreds and hundreds of them.

Here I am more in the heart of the oily dominions than elsewhere. I find that New Bedford and Nantucket, heretofore oildom, has been unsuccessful for years past, and is coming here with its millions of money and hordes of vessel officers, to harpoon the old mother of all whales (earth) and draw her blubber by the force of steam, which must eventually injure whaling oildom very much. Excitements are very common by greased excitements I never saw or heard of.

Drills going everywhere, steam-puffing, pump walking-beams, high above the buildings, working like those above a steamboat deck. Everywhere in sight barrels, mountain high, from steamers just in from Pittsburgh, towing boats and barges loaded with barrels. The common topic everywhere is oil, rock drilling, oil shows, depth, prices, prospects for oil for the future, rents of land, best sites, chances for rapidly made fortunes, what has been, and what will be.

My opinion is, that the landowners are grabbers; with the disposition to bless themselves and curse the world generally. They require one-half, and some as high as five-eighths, of the oil for ground rent; but time will settle all these things. The tariff is far too high, but men would prefer digging for the present, without a sure thing ahead, to prospecting at a greater risk. From ten to fifty barrels is the usual yield. Prices have ruled at $10 per barrel here, little less now, though I think when all its uses are found out, it will be much more in demand.

Enterprising, ambitious gentlemen rushed to the oil region. Some did well; others suffered misfortune.

HENRY R. ROUSE: PURSUING A DREAM

Henry R. Rouse came from Westfield, New York. He was born on August 24, 1824. His parents were Samuel and Sarah Rouse. Samuel seems to have left the household soon after Henry was born. Sarah was listed as head of household in the census data at that time. Henry spent most of his first sixteen years growing up in and around Westfield. The family lived on a farm on North Portage, located about halfway between Westfield and Barcelona Harbor. Henry attended school in Westfield until he was about twelve years old. Then he left Westfield, spending about two years at the Jamestown Academy.

Henry is reported to have "excelled in spirited orations." This caught the attention of Governor Seward. He made arrangements to place Henry in Abram Dixon's law office in Westfield to study law. Henry worked in the office for a few years before giving up law and the position. Contemporary sources suggest that there may have been a few reasons for his doing so. Some claimed that it was because of a minor speech impediment—Henry stuttered when talking to adults. Others claim a romantic issue ("being rejected by his love"). Whatever the reason, Henry decided to pursue another occupation.

Henry Rouse was about eighteen years old when he left Westfield in 1842. He moved to Warren County, Pennsylvania. He took up teaching for a time before opening a grocery and dry goods store in Enterprise, Pennsylvania, about five miles from Titusville. Henry then shifted over to the lumber business. He purchased land, invested in local businesses and managed to get elected to serve a few terms in Pennsylvania's General Assembly, the state legislature.

Henry was up for reelection when Colonel Drake struck oil. As Henry owned several parcels of land near Titusville, he took an interest in the oil business. William Barnsdale, a shoemaker in Titusville, lived just outside Titusville on the road to Enterprise. As Barnsdale's brother-in-law owned a farm near the Drake Well, Barnsdale decided to get in on the action by drilling a well. He didn't have the money, but Henry did.

Henry Rouse paid for the drilling tools. These were fashioned by a Titusville blacksmith. They hit oil in November 1859. The success of the Barnsdale Well was followed by more producing wells on Buchanan Flats. When wealth poured in, Henry became an oilman.

Forming an association with several investors, Henry hired Merrick & Hawley from Sherman, New York, to drill more wells. They were drilling near the Buchanan's Farm upper line on April 17, 1861, when Merrick and Hawley's crew struck oil. Henry wasn't there. He was sitting in the taproom at Anthony's Hotel in Titusville chatting with Mr. Perry, Mr. Buel and Mr. Barmore when George Hayes, a workman, rushed in to announce that they'd hit oil. George shouted, "It's monstrous; gotta get barrels."

The gentlemen rushed to the well. Henry approached the drilling shack to congratulate Mr. Page, his driller, when a great blast occurred. Fire engulfed the well. Flames leaped to a nearby well. Oil tanks, barrels and the Buchanan barn exploded. A sudden, roaring crescendo of flame towered in the air, waving, extending and growing. One poor figure stumbled from the inferno. Bystanders rushed to his aid. Wrapping his charred, naked body in a blanket, they carried him from the place. He died an agonizing death.

Nearby barrels burst into flame. A man standing near the barrels screamed in agony, burning to death. Henry Rouse stumbled from the blast, seeking safety in a nearby ravine. Stumbling a few steps, he fell into a circle of fire. Burying his face in mud so he could breathe, Henry recovered. Scrambling to his feet, he lurched forward a few steps toward the ravine. Falling again, Henry was unable to rise, unable to get up.

Two men rushed to his aid. Managing to endure the blazing heat long enough, they reached in, grabbed Henry's boots and pulled him from the flames. Henry was carried to a nearby shed and placed on a workman's bed. He gasped through five long hours of agony before dying. Henry's body, from the top of his head down his back and all the way to his legs, was burned to a crisp. He lived long enough to dictate his will before succumbing to his injuries.

Henry Rouse left the bulk of his $300,000 estate to Warren County, Pennsylvania. County officials built a $90,000 courthouse in Warren and appropriated the interest on the estate's remaining funds, for several years, to maintain Warren County bridges and highways.

Somewhat later, the *Westfield Republican* noted, "In a secluded spot in our village cemetery, with grave unmarked save by a small slab, lie the remains of Henry R. Rouse, a man whose princely munificence is benefitting thousands of people who have never known his name."

Warren County officials corrected the oversight. They purchased an appropriate grave marker to mark Henry's resting place. Warren County commissioners now visit and decorate Henry's grave in the Westfield Cemetery just before Memorial Day each year. An oil portrait of Henry Rouse hangs in the Warren County Courthouse, and another hangs in the Warren County Historical Society.

His legacy lives on in Warren County's Rouse estate. It's a specialized nursing and assisted-living facility in Youngsville, Pennsylvania.

PITHOLE: THE FIRST PIPELINE

A few log farmhouses stood along Pithole Creek in May 1865. The creek got its name from deep cracks and fissures in the creek bed. Underlying shale rock had been ripped and torn apart by ancient upheavals. The creek wasn't regarded as a particularly inviting place, as it gave off a strange, malodorous scent. Some claimed that it smelled like "sulfur and brimstone." Others believed that the creek was inhabited by the "Evil One." A few hardy souls ventured out to take a look at property along the creek, but there weren't many buyers.

The Holland Land Company decided to throw in an extra one hundred acres of land or so to sweeten the deal, but even then, there were few takers. Then Reverend Walter Holmden came along. He purchased two hundred acres near the headwaters of Pithole Creek in 1840 and moved his family to the site with the avowed intention of taking up farming. His son, Thomas Holmden, leased part of the property to Ian Frazer's United States Oil Company in 1864. Ian had already cleared about $250,000 as part owner of the Reed Well on Cherry Run; he was out and about looking for another opportunity.

Frazier hired a diviner and sent him out to take a look at Pithole Creek. When the diviner claimed that he "felt oil," Ian took a more active interest. He went to the creek and followed the diviner around until the forked twig dipped, pointing to a specific spot on the ground. Ian Frazier ordered his crew to drill on that spot.

The crew put up a crude derrick below the Holmden House that fall and went to work. Their first well turned out to be dry, but Ian Frazier continued drilling. He got oil on January 7. Petroleum shot into the air, drenching the derrick. The Frazer Well was the world's first gusher. It started out producing 650 barrels of oil a day, every day.

Kilgore and Keenan came along as Frazer was drilling. Taking an interest in the area, they formed a company that they named after themselves: Kilgore & Keenan. They started drilling two nearby wells. Their twin wells came in on January 17 and 19. The wells flowed at the rate of eight

This sign marks the entrance to Pithole, Pennsylvania. Pithole was an early oil boomtown. It boasted, for a few short days, the third-largest post office in the state. Although Pithole lasted for fewer than five hundred days, it gave the world its first successful oil pipeline. *Photograph by Hoffman.*

hundred barrels per day. The race was on. Drillers and speculators rushed to Pithole.

A Boston company soon arrived on site and started drilling another well, the Homestead Well. It got oil in April, getting five hundred barrels per day. The U.S. Petroleum Company subdivided its property and started selling lots for $3,000 for a half-acre plot. Thousands rushed to the area. Oil was discovered at two more wells in August: the Grant Well and Pool Well.

The *Titusville Herald* proclaimed Pithole as having "probably the most productive wells in the oil region of Pennsylvania." More than three thousand teamsters rushed to the area. Many were Confederate and Union war veterans seeking to profit from the opportunity.

Pithole City came into being, rising on the higher slopes along the creek. The first hotel, the two-story Astor House, was put up in a day, and by the end of the week, Holmden Street was boasting a teeming, thrown-together business center. The *Venango Spectator* noted, "Even while I write, buildings are going up, and some are put up and have groceries in them in six hours." Pithole was laid out in May, and by December, it had become an incorporated city boasting a population of about twenty thousand.

At its peak, Pithole, Pennsylvania, had fifty-four hotels, three churches, Pennsylvania's third-largest post office, a newspaper, a theater, a railroad and a red-light district "the like of Dodge City's."

Excitement continued. U.S. Petroleum drilled another well, which spewed five hundred barrels per day. Two days later, it got another well, and that one produced four hundred barrels per day. The company was making $2,000 per day. Then a storage tank burst; it dumped more than 2,500 gallons of oil into Pithole Creek. Two boys playing on the bank dropped a lighted match in the oil. It took more than two hundred men to put out the fire. "No Smoking" signs sprouted around town, but they did little good—this fire became the first of dozens.

Pithole was producing two thousand barrels of oil per day by the end of June, one-third of the region's total output. Long, winding processions of tank wagons traveled back and forth from Pithole to Titusville, Oil City and other shipping points. This heavy traffic soon broke up the roads leading into the town, destroying them. Conditions got so bad by July that many teamsters stopped hauling; they refused to do so.

Worried merchants and shippers in Titusville and nearby communities decided to raise a fund to fix the roads. After the teamsters were assessed one dollar per week to contribute to the fund, they raised their prices to three or four dollars per barrel for every barrel they hauled. As the price of crude oil

The Pithole Visitors' Center, set among the trees, is open for a few summer months each year. *Photograph by Hoffman.*

Former streets and building sites are still marked off on the hillside that was once Pithole, Pennsylvania. A model replica of the no longer existing community is displayed in the visitors' center. *Photograph by Hoffman.*

could fluctuate from as low as ten cents per barrel (during a surplus) to as high as ten dollars per barrel (during a shortage), producers weren't always able to cover hauling costs.

Seeking an alternative, Samuel Van Syckel, an oil buyer from Titusville, decided to try laying a pipeline. It would run from wells near Pithole to the railhead on Miller Farm, about five miles. Van Syckel borrowed more than $30,000 from the First National Bank of Titusville to fund his idea. He used some of his money to buy two-inch wrought-iron pipe in fifteen-foot lengths. The pipe was tested to withstand nine hundred barrels to the square inch. Joints were lap welded; each joint cost about $50.

Some of the pipeline was buried at a depth of up to two feet; the rest ran along the top of the ground. Once his pipeline was installed, Van Syckel purchased three Reed & Cogswell Steam Pumps. Two steam pumps were installed at Pithole and the other nearby. The pumps pushed crude oil through the pipeline, eighty-one barrels per hour, to the Miller Farm railhead.

As the pipeline took work away from the teamsters, they viewed it as an act of war, an assault on their livelihood and occupation. They had been able to disrupt past attempts at using pipelines to transport oil and figured that their usual tactics of sabotage and intimidation would work. But Samuel Van Syckel ruined their plans. He got carbines from New York and hired guards to patrol his pipeline. He issued orders: shoot on sight anyone attempting to damage the pipeline.

Van Syckel's pipeline worked, but he wasn't able to repay the loan. The First National Bank of Titusville foreclosed. It seized the property: pumps, pipeline and machinery. Jonathan Watson took over as manager and ran the line until oil production started dropping off.

The Frazer Well stopped flowing; then, one by one, other wells in and around Pithole started drying up, shutting down. There was a major fire in Pithole in February 1866 and another fire in May. This one burned eighty buildings, spread to thirty oil wells and burned twenty thousand barrels of oil.

Pithole's days were numbered. Buildings were taken down and carted off. A few people hung on in until 1867. From beginning to end, Pithole lasted for about five hundred days. However, Samuel Van Syckel gave the world its first successful pipeline.

Today, there is a visitors' center, open during the summer months. Otherwise, there are empty roadways. Signs mark spots where buildings once stood.

JOHN D. ROCKEFELLER: ORDER FROM CHAOS

John Davison Rockefeller was born on July 8, 1839, in Richford, New York. He was the second of six children. His father owned a farm and traded in many goods, including lumber and patent medicines. His mother, being opposite of his father in many ways, raised her family very strictly. The family moved to Oswego, New York, and lived there for several years before moving to Cleveland, Ohio.

John graduated from high school in Cleveland and attended Folsom's Commercial College for three months. Then he got an office assistant's job with Hewlett & Tuttle, commission merchants and produce shippers. When John was nineteen, he left Hewlett & Tuttle to create his own business, Clark & Rockefeller. Being commission merchants, it dealt in grain, hay, meats and other goods. It grossed $450,000 its first year.

Expanding the business, they purchased a small oil refinery in 1863. Wanting to secure a reliable source of oil on profitable terms, Rockefeller went to Franklin, Pennsylvania. He opened a small office by the French Creek Bridge and boarded at Anderson Dodd's place. Some who knew him at the time claimed, "No one ever accused John Rockefeller of not giving a hoot about making money." It was said that Rockefeller "could spot a good deal from a great distance."

Rockefeller wore an old, shabby suit of clothes around the oil field. He'd even pitch in and help load barrels when he was out and about. But come Sabbath, Rockefeller refused to have anything to do with business. He'd dress up in a new "Sunday go to Meeting" suit and hustle off to church. One Sunday morning, as Mr. Rockefeller was putting on his black frock coat, an employee rushed in to tell him that the river was rising and threatening to carry off their barrels of oil.

Several merchants, some oilmen and a bunch of others had already rushed to the riverbank, struggling to get their barrels to higher ground. Becoming anxious, the employee urged Mr. Rockefeller to come with him. However, Mr. Rockefeller put on his Sunday hat, saying, "Nope! I'm going to church." He went to church.

A lot of good oil got washed away that day, but what irritated everyone the most was that John Rockefeller didn't lose a single barrel. A few residents suggested that if this was how God rewarded Rockefeller for his Sabbath loyalty, God ought to keep a sharper eye on Rockefeller's business dealings during the week.

However, it was said that most folks got along with Mr. Rockefeller. He was actually held in considerable esteem by Mr. and Mrs. Dodd. They took in a bunch of boarders, but John Rockefeller was only boarder they ever had who just wanted bread and milk for supper.

Rockefeller came back to Cleveland to report, "Its chaos and disorder out there, waste and incompetence, competition at its worst." Deciding that refining was the place to be, he continued running his refinery. Rockefeller became senior partner. Clark was bought out, and the firm Rockefeller & Andrews grew to become Cleveland's largest refinery. During the 1866 depression, Rockefeller opened another refinery, the Standard Works. Then he opened a New York office to handle exports.

Henry Flagler came into the firm as Rockefeller's right-hand man, and by 1868, Rockefeller, Andrews and Flagler was operating the largest refinery in the world. Its size enabled Flagler to get a freight rebate of about fifteen cents per barrel out of the forty-two cents the railroads charged to bring a barrel of crude oil from the oil fields to Cleveland. Then investors started speculating.

A group of European bankers financed a ring in 1868 to try to get control of the oil export trade. American speculators jumped into the scheme. They started inflating the price of crude oil like a balloon but were ultimately overwhelmed by the supply. New wells kept coming on line. They flowed so lavishly that speculators couldn't keep up.

The fluctuating prices wreaked havoc with the oil industry. Then Jay Gould and Jim Fisk tried to corner gold. This spilled over into the oil industry when a group of Pittsburgh refiners teamed up with the Pennsylvania Railroad to raise $1 million to buy up enormous quantities of crude oil. The ring went about this so deviously that dozens of producers and agents signed contracts promising delivery without smelling a rat. The syndicate then got squeezed in a tight corner; it had to sell short, dumping 500,000 barrels of oil.

Then a number of producers tried several schemes to inflate the price of crude oil. This made the going difficult for small refiners. Lamp oil (kerosene) was their main product. The retail price for kerosene plummeted, dropping more than 50 percent during the five-year period between 1865 and 1870. Small refineries were hit hard—several closed. As Cleveland was getting most of the crude oil, Pittsburgh and New York raised a fuss.

Tom Scott, vice-president of the Pennsylvania Railroad, rushed to their rescue. He set up a fast new freight service linking Oil Creek with Philadelphia and New York. Rockefeller's new Standard Oil Company cut its own deal with the Erie Railroad. The *Cleveland Leader* noted, "Fortunately for us, the Erie Railroad managers have a keen eye on Mr. Scott and we may rely upon them to so adapt their oil rates to the market that Cleveland refiner interests will not suffer."

Standard Oil wasn't suffering, but John Rockefeller disliked competition. He devised a plan that he called "Our Plan" to consolidate refineries to reduce waste and avoid competition. Standard Oil being the biggest, it wouldn't get lost in the arrangement—it would lead the parade. As his first step, Rockefeller raised capital from local bankers to create a merger with other Cleveland oil refineries. After that, he tried to take on firms in other cities, tried to get the railroads to end their rate wars and tried to get the industry to regulate production, ending overproduction.

It was a good plan, but Peter Watson of Vanderbilt's Lake Shore Railroad came up with a different scheme with a benign-sounding name: the South Improvement Company. All three major railroads—Vanderbilt's New York Central, Tom Scott's Pennsylvania and Jay Gould's Erie—agreed to go along with the scheme. It was a ruthless plan. All three railroads would raise their rates between 100 percent and 250 percent. Refiners participating in the plan would receive enormous rebates on everything they shipped and also get the same rebate on every barrel their competitors shipped. The railroads promised to divide the share of profits equally.

Rockefeller, afterward, said that he didn't like the plan but went along with it. Twelve other refiners signed up. They got a few more to join them. But word got out on the arrangement, and the producers reacted by starting a boycott. They shut down the operation. However, Standard Oil made the most of the situation by absorbing twenty of the twenty-six refineries in Cleveland. Standard Oil got bigger.

John Rockefeller and Henry Flagler met with three Pittsburgh refiners in Titusville in May 1872 to form a voluntary alliance of refiners—no coercion, just friendly cooperative arrangement for the good of the industry. By the end of the summer, four-fifths of the firms in the United States belonged to the National Refiners Association. John Rockefeller was president.

Rockefeller and Standard Oil started examining other options. Intrigued by Van Syckel's success at conveying petroleum through a pipeline, Rockefeller decided that pipelines were the way to go. Pipelines moved oil, but more importantly, they determined where it ended up. John Rockefeller purchased United Pipe Line, a small group of lines, in 1882. Five years later, he controlled half of the existing lines. By the end of 1879, he controlled almost all of the pipelines in the United States.

As pipeline technology improved, Standard Oil started using pipelines to collect and convey natural gas. Several wells produced crude oil and natural gas. Some just produced natural gas. Collecting natural gas, conveying it through pipelines and selling it was a better business practice than igniting it and burning it at the wellhead.

Although some have questioned John Rockefeller's business practices, Standard Oil provided the emerging energy business with the stability, direction and coherence needed at the time. Standard Oil laid the groundwork for today's energy infrastructure, the same infrastructure that supports our world economy.

PART V

WESTERN NEW YORK OIL BUSINESS

JOB MOSES: LIMESTONE, NEW YORK

Limestone is now a small cluster of buildings situated on and along Route 219 in Cattaraugus County. It is located about two miles north of the Pennsylvania state line. The settlement is on Tuna Creek, a slow-moving stream that flows into the Allegheny River. The Senecas had a name for the creek. They called it the *Tunungant*, but newcomers found the name hard to pronounce. They called it the Tuna. Most of the early settlers came from New England, part of the New England exodus in the early 1800s.

Most were "wood choppers." Lumbering was their primary occupation. Each spring, the early settlers floated huge rafts of pine and hemlock logs down Tuna Creek to the Allegheny River. Then they continued down the Allegheny to Pittsburgh. If everything went well, they could get to Pittsburgh in a few weeks. It took them longer to get back. When they did, they continued clearing land, chopping down trees, doing a little subsistence farming and raising a few sheep. However, wolves were a problem—wolves liked mutton, and a few furtive panthers were noted as well.

The forest was cut back, and the trees were gone by 1860. Several Tuna Valley residents packed up and moved on, going west, but a few stayed behind. Then they got word coming up the river: Colonel Drake had struck oil near Titusville. A local search for oil started in 1864. James Nichols, Henry Renner and Daniel Smith leased one thousand acres on the Baillett

Tuna Creek still flows through Limestone, New York, on its way to join the Allegheny River. *Photograph by Hoffman.*

Farm near Limestone, and they started drilling. They got traces of oil at a depth of 570 feet. Continuing to drill, they got down to 600 feet before reaching the conclusion that they weren't going to find enough oil to make the effort worthwhile. They gave up and walked away.

However, word of their discovery got out. A group of New York City investors decided to pursue the venture. The investors bought the Hall Farm and several more acres around Limestone. Having purchased the land, they formed the Hall Farm Petroleum Company. The object of their company, as proclaimed in its articles of association, was to "mine and bore for the production of liquids and minerals, and to rend and sell the same."

Needing money to cover potential drilling costs, the investors contacted Job Moses, a merchant living in Rochester, New York. They negotiated an $8,000 loan. Job was an interesting man. He was born in Simsbury, Vermont, on August 30, 1815. His family moved down the river to Connecticut when he was a child. They left Connecticut the following year, going west, moving to Marcellus, New York. Job attended school in Marcellus, graduated and started teaching at the school. He taught for a number of years before becoming superintendent.

Then Job gave up his job and moved to Auburn, New York. He opened a drugstore and started selling patent medicine. His first experience with petroleum appears to have occurred when he became American Medicinal Oil's agent for New York, Michigan and the Canada West region. The medicinal oil was bottled in Burkesville, Kentucky. Advertised and sold as a cure-all for all sorts of afflictions, the medicine was crude oil. The oil was dressed up and sold in an imposing bottle. It had a fancy label and sold for one dollar per bottle.

Job left Auburn and moved to Rochester in 1856. After Job issued his loan to the New York investors to fund their Cattaraugus drilling operation, Job waited for a return on his investment. When that didn't happen, and the investors were not able to supply him with oil or repay his loan, Job took over the company, becoming president.

Having caught oil fever, Job put his medicine business aside, left Rochester and moved to Limestone. As he knew about the Cuba Oil Spring and was aware of oil discoveries occurring down the Allegheny River, Job came to the conclusion that there had to be a huge oil reservoir underlying the Allegheny Basin. Getting ready to test his conviction, Job figured that he might have to drill a bit deeper to get oil at the upper end of the Allegheny Basin, but he believed that it was there.

As the Hall Farm Petroleum Company already owned 1,250 acres three-quarters of a mile west of Limestone, Job looked it over and started drilling. He got oil at 1,060 feet. His first well put out two hundred barrels for part of the first day, but the well was lost by accident before its full value could be ascertained. However, Job was encouraged; he bought another 900 acres and leased another 1,000.

Job drilled his second well in 1867 a short distance west of his first well. He tubed this well to a depth of 1,100 feet, but another accident intervened. He put down his third well the following year. This well started out by producing ten barrels per day, but then it dropped off to two barrels per day. Nevertheless, Job continued. He put down his fourth well in 1871. He got oil at 540 feet and hit oil again at 1,100 feet. He didn't get much oil, but it was good quality.

Word got out. Harsh & Schreiber put down a well on William Beardsley's farm near the state line. At about the same time, Wing & Lockwood drilled a well on Hiram Beardsley's farm on the east side of Tuna Creek. It hit oil at 775 feet, producing about twenty-five barrels per day. Another well was drilled on the nearby Muller Farm by the Consolidate Land and Petroleum Company. That company got oil at 1,075 feet.

Inspired by the success of these companies, new ones came along and started drilling; 35-five wells were producing by 1875 and 225 by 1878. The largest-producing wells were the "Eureka" on the Clark Farm in 1877 and the "Irvine Farm Company Well" in the fall of 1878. Each well produced about 250 barrels per day.

However, the average yield of most wells in the Cattaraugus region turned out to be about ten barrels per day. Although this was not sufficient to generate the excitement that usually accompanied big "oil finds," the wells generated adequate revenue for those participating. Means had to be found to transport the crude oil to a refiner. Getting the oil to the nearest railroad was time consuming and expensive.

The Carrollton Branch of the Erie Railroad was the only line in the valley. It ran infrequent trains out of the valley. As the Erie had no competition, it charged producers about $150 per car. A branch of the Pennsylvania Railroad ran through nearby Olean, but the producers had to find a way to get their oil from Tuna Valley to Olean. The Pennsylvania Railroad solved this problem by hiring John H. Dilks to run a pipeline from Tuna Valley to Olean.

Dilks did this by laying a fourteen-and-a-half-mile, two-inch wrought-iron pipeline. Pipe joints were sealed using the new screw coupling. A pumping station was built at the foot of Rock City Mountain to pump the oil up the mountain. Once over the mountain, the oil flowed down the northern slope of the mountain to the Allegheny River. It crossed the river and flowed into storage tanks on Academy Hill in Olean. The storage tanks were forty feet above the Pennsylvania Railroad's tank car loading racks.

The *Olean Times* noted, "The elevation is double that of any other pipeline in the country and probably greater than any future company will be obliged to surmount." Crude oil entered the line at the Tuna Valley pump station on November 25, 1825, and flowed into the storage tanks in Olean two days later. The price of shipping oil out of Tuna Valley dropped from $150 per car to $100 per car. Then it dropped to $80 per car.

A sixty-barrel-per-day refinery was built on the McCarty Farm, about one mile from Limestone, New York, in 1878. In the course of oil development and exploration, several veins of salt water were discovered. As an interesting curiosity, several pieces of petrified wood were pulled from wells in the region. These wells were 185 and 200 feet deep.

Having profited from his Cattaraugus County endeavor, Job Moses decided to cross the state line and drill an exploratory well near Bradford, Pennsylvania. Forming a partnership with C.H. Foster and James Butts, they created the Foster Oil Company in 1871. Drilling at a point two miles northeast of Bradford, they got a ten-barrel-per-day output at 1,110 feet that November. Being the first successful well drilled on the Bradford field, the site of the "Uncle Job Moses Well" was marked by a memorial stone placed during Bradford's Diamond Jubilee.

Three years after the company's first oil strike, Butts & Foster drilled a second well. It got oil on the Buchanan Farm, half a mile northeast of its first well. This seventy-barrel-per-day producer came in on December 6, 1874. When word got out, speculators and producers rushed to Bradford. More wells were drilled, striking oil on the Webster lot and then the Watkins Farm, the Crooks Farm and, finally, the Tibbetts Farm. The Bradford boom was underway.

While this was happening, Job Moses built a twenty-two-room mansion at the end of his own street, Moses Street, in Limestone, New York. His house, built of clear white pine, was surrounded by a white picket fence. Job Moses took an active interest in local civic affairs, public education and politics while living in Limestone. He was elected town supervisor and served three terms as a member of the Cattaraugus County Board of Supervisors. Job Moses took the lead in consolidating three independent common school districts to form the Limestone Union Free School District, the first consolidated school district in Cattaraugus County.

This street sign marks Moses Street in Limestone, New York. The street honors Job Moses. His mansion once stood at the end of the street. *Photograph by Hoffman.*

The Penn Brad Oil Museum near Bradford, Pennsylvania, is located about ten miles south of Limestone, New York. The museum is marked by a seventy-two-foot-tall, circa 1890 drilling rig. Exhibits include information on Job Moses, drilling practices and legends of the world's first billion-dollar oil field. The Bradford Oil Field produced 83 percent of the country's total oil output in 1881. *Photograph by Hoffman.*

Job Moses lived in his white mansion at the end of his street until he purchased a home in New York City in 1877, moving there soon after—except for occasional visits to Limestone. His health was failing. After six years in New York, he decided to move back to Rochester to pass his remaining years with old friends. He passed away on July 25, 1887, ten years after leaving Limestone. His Limestone mansion is gone, but his street, Moses Street, is still there.

Job Moses had a strong impact on Western New York, Cattaraugus County and the oil business. Four thousand active wells on the Bradford Oil Field produced five thousand barrels of oil per day in 1880. The Bradford Oil Field produced 83 percent of the entire United States oil output in 1881. It became the world's first billion-dollar oil field.

OLEAN, NEW YORK:
EMBARKATION, PIPELINES AND REFINERIES

Major Adam Hoops, a Revolutionary War army officer, founded Olean in 1804. He selected the site because he believed that it was situated to become a great city—a thriving center of commerce, navigation and migration. He was right. Settlers started arriving early, many from New England. They came during the winter months, gathering on the banks of the Allegheny and waiting for the ice to break up. Tilly Buttrick, a Massachusetts native, wrote a description of the settlement and events taking place on the banks of the Allegheny when he came to Olean in 1815:

> It was much altered in appearance since my former visit here; instead of a few log huts as before. There were forty or fifty shanties, or temporary log houses, built up, and completely filled with men, women, and children, household furniture thrown up in piles; and a great number of horses, wagons, sleighs, etc.
>
> These people were emigrants from the eastern States, principally going down the Ohio River. Two gentlemen undertook to take a number of these people, and found it to be about twelve hundred people of all ages and sexes. They had a large number of flat-bottomed boats built for their conveyance; these were boarded up at the sides and roofs over them, with chimneys suitable for cooking, and were secure from the weather. There were also many rafts of boards and shingles, timber and saw logs, which would find a ready market at different places on the Ohio River.
>
> There are many saw-mills on the stream above this place, where these articles are manufactured from the fine timber which grows in vast quantities in this vicinity. The river at this time had risen to full bank, and I should suppose was navigable for vessels of fifty tons burden; but the river was frozen over.
>
> I waited about ten days, which brought it nearly to the close of March. On Saturday night I sat up late, heard some cracking of ice, several of us observing that we should soon be on our way to bed. Next morning at daylight found the river nearly clear, and at eight o'clock was completely so. The place now presented a curious sight; the men conveying their goods on board the boats and rafts, the women scolding, and children crying, some clothed, and some half clothed, all in haste, filled with anxiety, as if a few minutes were lost the passage would also be lost. By ten o'clock the whole

The Olean Historical and Preservation Society is housed in the Bartlett House at 302 Laurens Street in Olean, New York. The center contains genealogical, historic information and records pertaining to Olean, Cattaraugus County and surrounding region. *Photograph by Hoffman.*

river for one mile appeared to be one solid body of boats and rafts. What, but just before, appeared a considerable village now remained but a few solitary huts without occupants.

Olean's high tide as a migration center came in 1818, when more than three thousand settlers embarked on the spring floods going down the Allegheny; 350 people crowded on a single raft. Rafts and flatboats were usually guided by two or three men using long sweeps, tiller oars. When they reached their destination, rafts and flatboats were usually taken apart, and the lumber was used to build cabins.

Keelboats carried items up and down the river. They brought manufactured items up the Allegheny River from Pittsburgh and took back barrels of black salts, pearl ash and agricultural produce. The keelboats were from forty to eighty feet long and built like a ship with a timber keel. The eight- or ten-man crews walked along the deck, poling, pushing and pulling to get the keelboats upriver. Going upriver was a slow process, maybe making five or six miles in a day. Going downriver was better. Rowed by the crew and steered by a helmsman, keelboats could make thirty or forty miles per day. Keelboats operated on the Allegheny until the 1830s.

A few Olean residents took notice when Job Moses struck oil in nearby Tuna Valley in 1865. Even so, oil didn't have much of an impact on the Olean economy up until about 1874. That's when a few wells on both sides of the state line, in Cattaraugus and McKean Counties, started producing significant quantities of what came to be called "slush oil." The more productive "third sand oil" had not yet been reached. Most of the oil being produced at the time was in proximity to the Bradford branch of the Erie Railroad.

Heavier-quality oil then started to be sold to various parties for lubricating purposes. That was the situation when John H. Dilks came to Olean. He took a look at the region, checked out the situation and reached the conclusion, based on the proximity of railroads and geographic location, that Olean, New York, was positioned to become a major oil center. Dilks organized the Olean Petroleum Company, obtained rights-of-way and started digging a ditch, laying a 14.25-mile pipeline from Olean to Tuna Valley and Limestone.

He completed laying his two-inch wrought-iron pipeline in eighty days. It went over Rock City Mountain at an elevation of 2,350 feet, crossed the Allegheny River and came into Olean at the foot of Fourth Street, where the oil would be stored in storage tanks. Oil started to be pumped through New York State's first pipeline on Thanksgiving Day 1874. The line started out by moving a few hundred barrels per day, but within three years, the Dilks pipeline was bringing twenty thousand barrels of oil per day to Olean.

Dilks then laid a three-inch pipeline paralleling his first two-inch line. After that was done, he replaced his original two-inch line with a four-inch line. Dilks then sold his company to the Empire Transportation Company. Standard Oil came along and purchased the pipelines in 1877; it then extended and expanded the lines. Standard Oil pumped 175,000 barrels of oil into Olean in November 1878 alone.

Wing, Wilbur & Company decided to open a refinery in Olean in 1877. The company purchased a north Olean site and built a refinery with a refining capacity of six hundred barrels per day. It used Tuna Valley crude oil. A little more than a year later, Wing, Wilbur & Company sold its refinery to the Acme Oil Company of New York, a Standard Oil subsidiary.

The Acme Oil Company increased refining capacity by installing two more five-hundred-barrel stills. To ensure a constant supply and reserve of crude oil, Standard Oil put up three new thirty-five-thousand-barrel tanks. These became Olean's first storage tanks. Twelve new stills, new steam boilers, more tanks and other equipment were added to the Acme Refinery in 1879. Refining capacity doubled and then doubled again to more than three thousand barrels of oil per day. With its pipelines and railroads, Olean became the country's leading oil production center.

More than three hundred oil storage tanks surrounded Olean, New York, by 1880. Olean became the largest oil storage depot in the world. Standard Oil then made the decision to run a 315-mile pipeline from Olean, New York, to Bayonne, New Jersey. When this pipeline was completed in 1881, it was the longest petroleum pipeline in the world. The six-inch pipeline crossed fourteen rivers, twenty creeks and eighty mountain peaks on its way to New Jersey.

Workmen dug more than three hundred miles of trench for the pipeline by hand, using picks and shovels. Dynamite blasted short stretches through the mountains. Hundreds of teamsters hauled construction materials, often over rough and rugged roads chopped through dense forests. Four construction headquarters were set up, in Olean, Elmira and Binghamton, New York, and one in Passaic, New Jersey.

Each construction gang consisted of twenty-eight men, a telegraph operator and a time keeper. Every gang was expected to lay two hundred eighteen-foot joints of pipe every day. The pipe was lap-welded wrought iron with long, heavy collars. Ends and sockets were cut on a taper with nine threads to the inch, making a tight joint. The pipeline was buried eighteen inches deep.

Workmen strung telegraph lines over the route before they started laying pipe so that the company could keep in touch. It wanted to know what was going on while it was going on. Confidential messages were sent in code. Eleven pumping stations, twenty-eight miles apart, were constructed on the line. Every pumping station had duplicate boilers, engines and pumps so that the pumping station could continue pumping if it had a breakdown.

When finished, walkers patrolled the pipeline. They covered the twenty-eight miles between stations in two days. They patrolled the line on skis and snowshoes when there was deep snow. They had to keep the line open and operating.

By 1896, Olean's Acme Works had forty-nine operating stills, twenty-three steam boilers, fourteen steam engines, sixty-eight steam pumps, fourteen steam engines, sixty-eight steam pumps, 153 tanks of various sizes and endless miles of pipe. When in full operation, Standard Oil's Acme Oil Works refined five thousand barrels of oil per day, becoming the world's largest oil refinery.

NATURAL GAS COMES TO OLEAN: LIGHTING THE CITY

Drillers struck a roaring gas well on Indian Creek near Olean in 1877. Three years later, Joseph Pew, Justin Bradley, Edwin Bradley and Edward Emerson applied to bring natural gas to Olean. The city gave them an exclusive franchise to lay pipe in the streets. The agreement required building a gasometer in sixty days. For the plant to be operating by August 1, 1881, gas had to be high quality, and the price could not to exceed $2.50 per one thousand metered cubic feet.

With franchise in hand, the founders applied to the State of Pennsylvania for a charter for the Keystone Gas Company. Governor Hoyt signed the Letters Patent or Charter for the company on February 8, 1881. The stated purpose of the Keystone Gas Company was to supply natural gas for drilling and pumping wells along the summit of the mountains from the New York/ Pennsylvania state line, near the head of Indian Creek, to Big Shanty.

The company then contracted with the City of Olean to supply natural gas to fifty streetlamps and to "keep the lamps lighted from sunset to dawn for two years from January 1, 1884, at a cost of two dollars a year for each lamp so lighted." The rate for lighting these streetlamps went up from two dollars per year for each lamp to six dollars per year in January 1888.

Natural gas flowed into Olean. As time went on, it appeared that more gas was flowing into the city than what was being recorded by individual meters. Therefore, the Keystone Gas Company sent inspectors into Olean to check every building in the city during the summer of 1901. They were directed to carefully check all of the company's lines and meters. Auditors checked the gas bills paid by each consumer for the previous year. In some instances, the inspectors found elaborate piping systems that bypassed the meters. It appeared that several customers were handy with their own wrenches and fittings.

Inspectors also discovered that a few meters that had been turned upside down, which caused them to fail to register gas flowing through them. As a result, a number of customers were called to the company office to explain these irregularities. Those who failed to provide adequate explanations were presented with bills of $100 or more to correct calculated deficiencies. A follow-up inspection four years later revealed no more tampering with gas meters or their connections.

Olean consumers were supplied with gas from company-owned wells from 1881 to 1885. Starting in 1885, additional gas was purchased from the United Natural Gas Company. The directors of the Keystone Gas Company had the good fortune of selecting Olean as their market for natural gas because the population of the city increased from about three thousand people in 1880 to more than twenty thousand residents over the course of the next few years.

Keystone started bringing natural gas from West Virginia into Olean in 1940. Olean, New York, was later connected to Texas and Louisiana by more than 1,565 miles of natural gas pipeline.

Allegany Oil Fields: Scio, Bolivar and Richburg, New York

I remember traveling through Bolivar, the Allegany Oil Fields and on to Wellsville, New York, to visit Maude Jones, a relative, in the late 1940s. Maude's husband, Ralph Jones, was an old self-proclaimed "oilman." He told fascinating yarns about pipelines, drilling disasters and other things that happened on the Allegany Oil Fields in the old days.

It took us a little over four hours to drive from Chautauqua County to Maude's house in Allegany County. Dad usually drove through South Dayton, New York, to New Albion and on to Salamanca. Once we reached Route 17, we took the highway through Allegany to Olean and up the main street of Olean. Then it was on to Portville and Bolivar. You could smell it, the oil, after you got through Portville. The smell wafted in through the open windows of the car. There were pumps and pumping stations all along the way. Central pumping stations had miles of cable stretching from the engine house out across the fields to individual pumps, rocking up and down at each well.

Long cables creaked back and forth, operating the pumps. Oil wells were scattered all across the landscape, pumping beams rocking in unison. Tangles

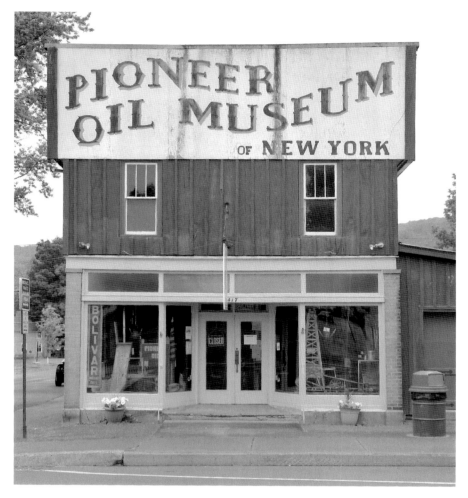

The Pioneer Oil Museum is located in what used to be the McEwen brothers' oil field supply store in Bolivar, New York. The museum's exhibits, collection of oil field equipment, records and displays tell the story of ones of the nation's oldest oil fields: the Allegany Field, extending from northwestern Pennsylvania into New York State's Cattaraugus and Allegany Counties. *Photograph by Hoffman.*

of pipe and cable were scattered about. The oil fields went on for miles and miles. Most wells were far out, away from the road, but from time to time, there'd be an oil well next to the road or in someone's front yard.

The Allegany Oil Fields were impressive, but now they're history. However, Bolivar, and the rest of southwestern New York, is sitting on a huge bed of

Marcellus shale. The United States Geological Survey has estimated that this shale contains more than 1.9 trillion feet of natural gas, enough to last for at least one hundred years. Natural gas is being recovered from Marcellus shale in nearby Pennsylvania, but regulatory issues are waiting to be resolved in New York State.

Orville P. Taylor launched the Allegany County Oil Industry with his celebrated "Triangle No. 1 Well." The well was drilled on the Crandall Farm near Scio, Allegany County, on June 12, 1879. It was a flowing well. The *Elmira Gazette* reported the event by announcing that "oil has been found at Wellsville, in paying quantities. Triangle well, at four and a half miles from Wellsville, was commenced on April 17th. They reached oil bearing sand at 1,100 feet. They stopped drilling 1,117 feet. The well was torpedoed and oil was sent flying twenty to thirty feet over the derrick. The well has been flowing between eight and ten barrels a day. The place has been named Triangle City."

Taylor's Triangle City gusher launched a frantic rush to Allegany County. Orville Taylor had been making and selling cigars in Wellsville before going into the oil business. He put down dry two wells before striking oil with his Triangle No. 1.

Men started drilling in Allegany County. On the morning of April 27, 1881, another gusher came in a mile west of the crossroads community of Richburg. Richburg had been a sleepy little rural town with a few dozen houses, but that soon changed. When word got out, oil scouts rushed to Richburg; the telegraph carried news of the gusher to newspapers in all parts of the country. Thousands read the news, and the invasion started. Richburg suddenly turned into the wildest and most wicked boomtown in the United States.

Oilmen poured across the state line, from Bradford, Pittsburgh and Oil City. They came from all points of the compass, bringing the unsavory army of camp followers that followed oil discoveries in those days. There was a scramble for leases. In a week, the population of Richburg grew from a hamlet of two hundred people to a veritable city of eight thousand. Four stage lines started operating. Big old-fashioned coaches, pulled by four horses, lumbered into Richburg, laden with oil-crazed passengers. Flimsy houses and stores with false fronts sprang up.

There were few sleeping accommodations. It was nothing strange to see twenty men crawl out of a haymow at daybreak. Sometimes as many as two hundred men slept under the maples of the little park by the village schoolhouse. Others paid one dollar per night for the privilege of sleeping

Above: This old Buffalo Oil Field gasoline engine is proudly displayed in the Pioneer Oil Museum. Buffalo engines were once manufactured by the Buffalo Gasoline Motor Company, which was located at 1280–90 Niagara Street in Buffalo, New York. *Photograph by Hoffman.*

Left: Kelley Lounsberry, director of the Pioneer Oil Museum, as he demonstrates and explains the use of old drilling equipment in the museum's replica of an "old drilling shack." *Photograph by Hoffman.*

on a billiard table. One saloonkeeper did not wait for carpenters. He placed two whisky barrels on their ends, used a plank for a bar and set up business in the street.

At one time, there were 135 women plying their peculiar profession in the town. There were gambling joints in barns and over stores. Money flowed like water. There were murders, stabbings and gang fights. No mining camp in the Rockies was wilder than Richburg at the height of its oil boom.

The Allegany Central, a narrow-gauge railroad, was quickly constructed over the hill to link Richburg with Friendship and the rest of the world. The line was then extended to Olean. An empty freight car served as its station for the first month. A few weeks later, the Bradford, Eldred & Cuba Railroad arrived. It ran from Bradford to Bolivar, crossing the Allegany Oil Fields to Wellsville.

An elaborate opera house also opened on Main Street. Stars of the spoken stage came to Richburg as part of the oil circuit. John L. Sullivan came there on one of his barnstorming tours. "Oil dippers" flourished for a time when the overflow of oil formed veritable ponds in low places. Two men threw up a dam on the flats. For a time, they were skimming and storing seven hundred barrels of oil. Another man set up a tank near a creek coated with floating oil. Oil-laden water flowed into his tank. A horde of men, boys and women spent their days dipping oil off the water, wherever it accumulated.

Farmers became rich overnight if oil was struck on their land. Men who had never seen more than fifty dollars at any one time in their lives now counted their wealth in the thousands of dollars. But Richburg's days of glory were short-lived. Word arrived of a new oil discovery, a gusher at Cherry Grove, Pennsylvania, in 1882. Almost as quickly as they had come, the fickle army that followed the oil fields packed up and moved on.

Even so, oil production continued around nearby Bolivar up until the 1950s before dropping off. Bolivar thrived during the oil days. Mementos, equipment, photos and exploits are celebrated in Bolivar's Pioneer Oil Museum.

However, Wellsville's huge Sinclair Refinery continued operating as long as the network of pipelines brought crude oil from nearby oil fields. The oil is gone, but Southern Allegany County continues to be one of New York State's most reliable sources of natural gas. In the early days, natural gas was considered more of a hazard than an asset since there was no practical way of capturing it, storing it or transporting it.

ALLEGANY COUNTY REFINERIES: FRIENDSHIP, BOLIVAR AND WELLSVILLE, NEW YORK

Allegany County's first oil refinery was built at Friendship, New York, in May 1880 by Charles Davis of Titusville, Pennsylvania, and William Flint of Friendship. It was a small lubricating works with a ten-barrel still operated by Mr. Davis and an assistant. This first refinery caught fire and burned down soon after opening. It was replaced by a new, larger refinery.

This second refinery was constructed by the Friendship Refining Company. It was built on a switch of a newly completed narrow-gauge railroad linking the newly discovered Richburg Oil Field with the rest of the world. The railroad hauled crude oil from the oil field in two twenty-barrel tanks set on a narrow-gauge flatcar. The plant was refining eight hundred barrels per month by June 1883. Business was blooming.

The refinery soon purchased the entire production of a wildcat well drilled on the Elisha Farm near Friendship. The oil was hauled by a team driven by Thomas Hamilton. He was paid $3.50 for delivering a load of five barrels. Barrels were filled at the well and emptied at the refinery.

Mr. Davis and his associates sold their refinery to Rigdon & Spears in 1883. One of the new owners, John W. Rigdon, happened to be the son of Sidney Rigdon, a noted Mormon leader. Sidney came to Friendship in 1847 after having had a dispute with Brigham Young. John was a successful attorney in Friendship, getting involved in the oil business. It turned out to be a good business decision; the Friendship Refinery prospered, making lamp oil, benzene, lubricating oil and white Vaseline.

Neighbors, however, saw things differently. They complained about the plant's offensive odor. Some threatened to burn it down, especially when the wind came from the east or southeast. One night, in 1884, the inevitable happened: the refinery and its products did indeed burn to the ground.

B.W. Baum & Son, dealers in oil leases, took action. It built the Cynthia Oil Works in Bolivar in 1882. This refinery was designed to manufacture lubricating oil. It had a capacity of eighty-five barrels per day. Baum & Son had nearly ten thousand barrels of crude oil in stock when it opened its refinery; much of the oil had been dipped or salvaged from the Little Genesee Creek near the works—it was overflow from tanks on the oil field. The refinery turned a profit so long as it could dip crude oil. When it wasn't available, Baum & Son shut down the plant.

The Wellsville Refining Company opened a modern refinery in Wellsville, New York, on September 27, 1902. This refinery opened with two six-hundred-barrel horizontal stills and sixty thousand barrels of crude oil. The Philadelphia and Reading Railroad built fifty tank cars with a capacity of 12,600 gallons each to transport the refinery's products. A two-inch, seventeen-mile pipe line was laid from the refinery to West Bingham, Pennsylvania, where illuminating oil was delivered to Marcus Hook, Pennsylvania. Most of it was shipped abroad.

Sinclair Consolidated Oil Company bought a 20 percent interest in the Wellsville Oil Refinery in 1919 and purchased the remaining 80 percent in 1920. Sinclair then increased the size of the refinery. By 1932, the plant covered 132 acres of land and employed more than five hundred people. It paid about one-fifth of the local school and town taxes.

Reaching its peak in 1937, the Sinclair Refinery operated for twenty-four hours per day. The refinery processed more than eight thousand barrels of crude oil per day and employed more than six hundred workers. At the time, George Tabor Jr., the directing head of Sinclair Refinery, described its Wellsville Refinery as "the largest and most modern refinery processing Pennsylvania grade crude."

The Wellsville Refinery survived a major fire in 1938 when a naptha tank exploded. The fire killed three refinery workers and injured another forty-two people. It took firefighters two days to put out the million-dollar fire. The refinery recovered to play a major role during World War II, supplying petroleum products to support the nation's war effort.

When a second major fire occurred in 1958, Sinclair closed the Wellsville Refinery. The Educational Foundation of Alfred State College has since taken over much of the site and is occupying the remaining buildings.

PART VI

ELECTRICITY: CREATING LIGHT AND ENERGY

CHAUTAUQUA AND ELECTRIC LIGHTS: THE CHAUTAUQUA INSTITUTION

Daniel Boorstin claimed, "The striking feature of New England public life was its abundance of reform movements." New England transplants brought this feature along with them when they came to Western New York. This reform imperative was exemplified, and still is, by the people, programs and presence of the Chautauqua Institution. It's located on Chautauqua Lake in Chautauqua County in Western New York.

Today's Chautauqua Institution consists of more than 1,200 Victorian cottages, contemporary homes, condominiums, hotels (big and small), shops and public meeting halls. It covers 225 acres of gated community on the shore of Chautauqua Lake. The Chautauqua Institution was established in 1874 by Dr. John Heyl Vincent, a Methodist Episcopal minister, and Lewis Miller, an Akron, Ohio inventor. It started out as a summer training program for Sunday school teachers. President Grant visited Chautauqua in 1875.

Since then, nine other presidents have visited Chautauqua during their term in office. When President Theodore Roosevelt lectured at the institution in 1905, he proclaimed, "Chautauqua is the most American thing in America." President Franklin Roosevelt gave his "I Hate War" speech in the Amphitheater at Chautauqua in 1936. President Bill Clinton spent a week at the Chautauqua Institution in 1996.

The Chautauqua Institution created the world's first correspondence program in 1878: the Chautauqua Literary and Scientific Circle. The program was designed to promote habits of reading and study in nature, art, science and literature. The CLSC launched the Chautauqua Movement.

About three hundred Chautauqua Assemblies were operating across the country by 1904. The national Chautauqua Circuit flourished between 1910 and the late 1920s before giving way to film, radio and, eventually, television. The original Chautauqua, on Chautauqua Lake, survived, growing and expanding along the lakeshore. George Gershwin composed his "Concerto in F" in a practice cottage at the Chautauqua Institution in 1925. Family members have attended Chautauqua for five or six generations. I spend a few weeks there each summer.

Electric lighting appeared at the Chautauqua Institution in 1877. This occurred when Lewis Miller invited his friend Charles Brush to demonstrate his new electric machine at the assembly. Charles Francis Brush, an Ohio native, was a graduate of Cleveland High School and the University of Michigan. He was working as an analytical chemist when he started working on developing an improved arc lighting system. Needing a reliable source of power, he improved on the work of others by developing his improved dynamo.

His dynamo was simpler than previous machines, but it generated a stronger current and was easier to maintain. First assembled in 1876, the dynamo was demonstrated at the Chautauqua Assembly in 1877. It was amazing. Arc lights illuminated the auditorium, providing enough light "to be able to read a hymnal during evening services." Thomas Edison, a Chautauqua attendee and son-in-law of Lewis Miller, took note.

Although Charles Brush was relatively unknown, he submitted his dynamo for an evaluation by the Franklin Institute of Philadelphia in 1878. The institute's evaluation committee tested four dynamos. The Brush Dynamo was selected as the most efficient, most powerful and easiest to maintain. The Franklin Institute effectively endorsed the Brush Dynamo as the machine of choice for generating electricity.

Chautauqua already knew this; arc lights had been installed in the auditorium and on the grounds. As the site was still a campground, attendees had to make do with tents and a few board-and-batten cottages. Tents were perched on wooden platforms clustered around a grove. There was a "hotel" on the grounds. It was an open, two-story framework shed; interior canvas walls divided off "rooms."

The architectural style of the Athenaeum Hotel at Chautauqua is similar to other "Saratoga style" resorts built during the later years of the nineteenth century. However, the building is unique in that it was specifically designed and wired to use electricity. *Photograph by Hoffman.*

As attendance increased, Lewis Miller came to the conclusion that the grounds needed bigger and better accommodations. He hired architect W.W. Carlin to design an appropriate, modern hotel for the Chautauqua Institution. Legend has it that ninety men put up the original Athenaeum Hotel in ninety days at a total cost of $125,000.

The Athenaeum opened in 1881. Electric wiring and lighting were installed as the building was built, making it one of the first hotels in the world to be wired and lit by electricity. The builders used Thomas Edison's direct current, along with his new incandescent bulbs. A few guests expressed concern about sleeping inside a building wired with electricity. Some didn't believe that it was safe, so they slept outside in tents. The grand old hotel is still standing. It overlooks Chautauqua Lake, and it's being used. It's busy every season.

With its increasing demand for electricity, the Chautauqua Assembly purchased a larger, more powerful generator from the Cleveland Electric Company in 1887. The generator, its boiler and a twenty-horsepower steam engine were installed near the Athenaeum Hotel. Sixteen arc lights were placed around the grounds, in the auditorium, the pavilion and over Palestine Park. The *Assembly Herald* claimed, "The electric lights made the

This scene is derived from an old stereoscopic photo of the front porch of the Athenaeum Hotel. It was taken during the 1880s. *Courtesy of Chautauqua Institution, prepared by Dening.*

Thomas Edison and Henry Ford seated together on a bench at the Chautauqua Institution. The men were friends and spent time with each other at the institution during the summer seasons. *Courtesy of Chautauqua Institution, prepared by Dening.*

grounds so brilliant that a person can read the Bible on even the darkest of nights."

Electricity was then extended to nearby cottages within two hundred feet of the generator. The entire Miller Park section of the Chautauqua Institution was illuminated during the 1890 season. Electric lighting became commonplace at Chautauqua several years before Buffalo, New York, gained fame as the "City of Light" during its 1901 Pan-American Exposition.

Buffalo's exposition and the Chautauqua Institution were both lit by electricity, but they used different systems. Chautauqua used Edison's direct current. It was produced by a steam engine–powered generator located on the Chautauqua grounds. Edison favored direct current; he had more than 1,500 local direct current power stations operating by the late 1890s.

However, direct current transmission was limited. A direct current power station could only provide power for up to about one square mile. It could light a city block, but the next block would need another power station. Edison lit a block with direct current in New York City in 1882.

Buffalo took a different approach. It used Nikola Tesla's alternating current.

ADAM, MELDRUM AND ANDERSON:
ALTERNATING CURRENT

The Adam, Meldrum and Anderson Department Store was housed in a gigantic, five-story Italianate structure located on lower Main Street in Buffalo, New York. It became the world's first commercial enterprise to be lit by electric lights. This happened in 1886. Buffalo was a bustling, growing city at the time. It was becoming rich as the transit point for immigrants going west and products coming east on the lakes. The waterfront was crowded with shipping and lined with grain elevators. More than twenty different railroads converged on the city.

A front-page ad in the November 17, 1886 edition of the *Buffalo Commercial Advertiser* announced the event: Adam, Meldrum and Anderson would be displaying its incredible selection of merchandise using 498 electric lights powered by the Westinghouse system. The ad claimed, "No odor, no heat, no matches; no danger."

Two days later, on Monday evening, November 19, the AM&A Department Store opened for business—not necessarily to sell merchandise but rather to show off its lights. Well-dressed crowds streamed up and down all four floors to marvel at the light, exclaiming how it was brighter than sunlight, talking about how you could even see the colors in shawls and the weave in draperies. The *Advertiser* reported, "The store was so thronged with visitors that it was difficult to get about."

The newspaper article failed to point out or explain the difference between the new Westinghouse system using alternating current and Edison's direct current. George Westinghouse opened his first commercial power station in Buffalo in 1886. Located in the Brush Electric Light plant at Wilkeson and Mohawk Street, power was produced by a steam-powered Westinghouse 400 lamp, single-phase, one-thousand-volt generator.

Alternating current could be transmitted for a considerable distance before being stepped down for safe consumer use. Generating stations could now be located some distance away from cities, closer to a fuel or power source. Alternating current could be manufactured and dispatched by power grids.

Niagara beckoned.

POWER FROM NIAGARA: CREATING THE POWER GRID

The spectacular grandeur and lure of Niagara Falls intrigued early viewers. Aaron Burr fancied placing the projected capital of his contemplated North American empire at the falls. He sent Theodosia Burr, his daughter, into the wilderness of Western New York to view the falls on her honeymoon in 1801. This started a tradition, and Niagara Falls became the honeymoon capital of the world. Others came, but they had more prosaic interests. Businessmen figured that there had to be some way to put the falls to work, harnessing the power of falling water.

The Niagara Falls Hydraulic Power & Manufacturing Company was chartered in 1853. The company purchased water rights and started digging a thirty-five-foot-wide, eight-foot-deep canal to divert water from the river above the falls in 1860. The canal was completed in 1861 and used to power a flour mill.

Jacob Schoelikopf purchased the canal in 1877 and used it to power his Chemical and Dye Company. Power was transmitted from turbines by using belts and drive shafts. Electricity was in its infancy, used only for the telegraph and newly invented telephone. However, the falls were being illuminated nightly by calcium flares as a tourist attraction. However, the flares were expensive and didn't last that long.

Charles Francis Brush, developer of the electric arc light, came to the falls in 1881 to install sixteen of his electric carbon arc lights and a generator to illuminate the falls. Schoelikopf offered power from his water turbines to power Brush's generator. The generator produced direct current. Direct current lit the arc lights, but it could only be transmitted for about two miles.

A group of local businessmen formed the Niagara River Hydraulic Tunnel Power & Sewer Company in 1886 and secured a charter to divert water from the upper Niagara River, but construction was stalled. The company was reorganized in 1899 to become the Niagara Falls Power Company. The new company was backed by several influential New York City financiers: John Jacob Astor, William K. Vanderbilt and J. Pierpont Morgan. Edward Dean Adams, a New York City businessman, served as president. Uncertain as to how to go about creating and transmitting power from the falls, the company offered a $100,000 prize for anyone who could come up with a solution. Nobody responded.

Two years went by before Westinghouse submitted a proposal using alternating current; this proposal was based on work done by William Stanley

Falling water at Niagara intrigued early visitors. Many figured that there had to be some way to harness the power. Millraces and canals eventually gave way to turbines and generators. *Photograph by Hoffman.*

and Nikola Tesla. William Stanley, an electrical engineer and innovator, was born in Brooklyn, New York. He attended Yale for a time before taking a job in the emerging field of electricity. He designed one of the country's first electrical installations for a New York City store before going to work for Westinghouse. Stanley is known for inventing the induction coil, which this made it possible to vary the voltage of alternating current.

Tesla came from Serbia. He studied engineering at the University of Graz in Austria before inventing his alternating current induction motor. He took a position with the Edison Electric Company in Paris before moving to the United States. He worked with Edison for a while before leaving to start his own firm. Then he joined Westinghouse. Tesla helped design the alternating current used at the Chicago World's Fair: Columbian Exposition. He put on spectacular electric demonstrations after the fair opened in 1893—Tesla caused metal eggs to rotate within polyphase electrical fields, he illuminated eerie neon lights and he created glistening tongue-like flames of electricity that issued from his generators and coils.

One fascinated observer, Colonel Henry Prout, wrote that "the best result of the Columbian Exposition of 1893 was that it removed the last doubt of the usefulness to mankind of the polyphase alternating current. The conclusive demonstration at Niagara was yet to be made, but the World's Fair clinched the fact that it would be made, and so it marked an epoch in industrial history."

The Niagara Power Commission reviewed several proposals. Thomas Edison and William Kelvin argued against the Westinghouse proposal; they claimed that high-voltage alternating current was too dangerous. The commission finally reached a decision in late October 1893, accepting the Westinghouse concept to generate alternating current, put the equivalent of 100,000 horsepower, generated by electricity, on a wire and send it to Buffalo. Construction started on Power House No. 1, the world's first major hydroelectric power plant.

Power House No. 1, the Adams Station, was constructed in layered sections. It was highlighted by huge, circular-topped windows measuring 14 feet wide and 15 feet high. The powerhouse building was originally 140 feet long but was extended to 450 feet when all of the generators were installed. Water that went *whooshing* into the station generated more than 100,000 horsepower.

Dropping into 8-foot-wide pen stocks (huge pipes), the water plunged straight down, more than 100 feet, around crooked elbows, roaring into the double blades of gigantic twenty-nine-ton turbines, the world's largest.

The "Arch Entrance" to the original 1895 powerhouse has been reconstructed on Goat Island in Niagara Falls, New York, as a testimony and memorial to Nikola Tesla. *Photograph by Hoffman.*

Turning the turbines, water rushed through a 6,800-foot tailrace tunnel, back to join the river below the falls. The turbines powered huge Tesla generators; the generators produced electricity.

Adams invited Nikola Tesla to speak at the opening ceremonies on January 27, 1897. Stepping to the podium, Tesla received a "monstrous ovation." Guests sprang to their feet, waving their napkins, cheering the world's greatest electrician. It took two or three minutes before quiet prevailed. Then the enigmatic electrical wizard started his talk in a subdued fashion, warming up to talk about the contributions of Bell, Edison, Thomson, Westinghouse and others to the enterprise.

Tesla was planning a rousing finale, one that might have prompted a few boos from some in the audience. He wanted to announce that the Niagara project was already obsolete; he had a batter plan—transmitting power without wires. However, the master of ceremonies, Mr. Stetson, had already read the speech. Stetson interrupted at that point, whispering in Tesla's ear. "I am just informed," Tesla told his audience, "that in three minutes we have to leave."

"No," the audience cried.

"Buffalonians," Tesla concluded, "I would say friends, let me congratulate the courageous pioneers who embarked on this enterprise and carried it to success." Tesla left the stage.

Once it had power, the Niagara Falls Power Company contracted with the Cataract Power and Conduit Company to construct a long-distance electric transmission line from Niagara Falls to Buffalo, twenty miles away. The company went to work, putting up cedar poles and stringing wire capable of carrying the eleven thousand volts being generated at the falls. Transformers were designed, built and provided by Westinghouse, creating a power grid— the world's first. Buffalo came on the new grid on November 16, 1896. Power had been transmitted to Buffalo. Now it needed a demonstration.

PAN-AMERICAN EXPOSITION: CITY OF LIGHT

Optimism, growth and excitement characterized Buffalo, New York, in 1901. It had become the nation's eighth-largest city and was a thriving center of commerce, shipping and industry. The new Ellicott Square Building was the world's largest office building. Buffalo enjoyed an elaborate public park system designed by Frederick Olmsted, and the city had more miles of paved highway than any other in the world.

Moreover, Buffalo had power—lots of power, power to spare. Electricity was brought to Buffalo on transmission lines from Niagara Falls. Assuming its role as "Queen City of the Great Lakes" Buffalo sponsored and organized the Pan-American Exposition. The exposition was designed to mark Buffalo's

The Buffalo and Erie County Historical Center is located on Museum Court in Buffalo, New York. The building was constructed in 1901 as the New York State pavilion for the Pan-American Exposition. It's the sole surviving permanent structure left from the exposition. *Photograph by Hoffman.*

emerging role on the national and world stage while celebrating the impact and significance of electricity.

The intent was to build an exposition more beautiful than any the world had ever seen: placid pools, dancing fountains, architecture embellished with sculpture and illumination during the evening hours by the marvel of electricity. Although much of the overall design for buildings and structures was derived from Latin American/Spanish influences, designers drew on Italian and other European influences to create a hybrid style that they called "Free Renaissance."

Buildings, with the exception of the New York State Building, were thrown up. Most were made by spreading a mixture of plaster and fiber, like stucco, over wooden frames. Painted bright hues of brilliant color, the buildings were striking but ephemeral. The bright colors enchanted some but offended others. The Pan-American Exposition—notable for vivid color, novel decorative architectural schemes and boldness of design—became known as Buffalo's Rainbow City.

The Rainbow City was a thirty-minute trolley ride from downtown Buffalo. Upon entering the grounds, the visitors were treated to the sight of splendid domes, attractive minarets, towers and pavilions gleaming with pleasing hues, tints and regal statues. Buildings contained exhibits from every part of the Western Hemisphere. The gates opened on May 1, 1901.

The general plan of the grounds was that of an inverted T, with the cross arms being the Esplanade extending east and west. It terminated at the Propylaea. The Court of Fountains was in the center of this vertical stem, and starting from its four corners was the beginning of the main group of large buildings.

The buildings were colored in various hues of brilliant red, blue, green and gold. The dominant feature of the exposition was the Electric Tower. The tower had a steel frame, rising 391 feet above the grounds. It could be seen from downtown Buffalo. During the daylight, the tower was a deep green, decorated with details of cream, white, blue and gold.

A spiral stairway in the center of the Electric Tower led up to a domed cupola housing a superb figure, the Goddess of Light. An electric elevator carried visitors to many floors. A large restaurant, seventy-five feet above the grounds, provided a superb bird's-eye view of the entire grounds with its colonnades and festivities.

Most spectacular was the evening. The grounds were bathed in the light of 240,000 electric light bulbs lit by electricity brought from Niagara Falls. The Electric Tower alone was studded with 44,000 lights. A powerful searchlight mounted at the top of the tower roamed back and forth over the grounds, illuminating them. The searchlight could be seen from Canada and Niagara Falls.

Thousands came to the grounds, including President McKinley. Unfortunately, President McKinley was shot by an assassin, Leon Czolgosz, at the reception held on his behalf in the Hall of Music on September 6, 1901. Attendance dropped off at the exposition after the assassination. However, thousands of people were awed by the brilliance of electric lighting. It became the desired standard. Although kerosene lighting persisted for some in some areas, its lure was lost.

Alternating current could transmit electricity for great distances, and electrical power became the norm. Cities and communities throughout the United States rushed to get on the expanding electrical grid. Many came to enjoy the convenience of electrical power and lighting in the course of the next few years.

As people rushed to get on the grid, the market for lamp oil dried up. Petroleum producers faced financial ruin, but inventors rushed to their rescue. Edward Butler, a British engineer, developed a gasoline-powered internal combustion engine in 1884. He created the spark plug, the magneto, a coil ignition system and the carburetor—all are required to make the engine run. Karl Friedrich Benz put an internal combustion engine in a carriage in

1885. Then Charles and Frank Duryea built a gasoline-powered automobile in Springfield, Massachusetts, in 1893.

Francis and Freelan Stanley took a different approach. They burned gasoline to create steam and used the steam to power the engine. They started building gasoline-burning Locomobiles at their Bridgeport, Connecticut factory in 1898. Then they diversified the product by making three models: Stanhope, Locoracer and Locosurrey.

Perceptions of distance changed when people discovered they could one, use fossil fuel to power a generator; two, use this generator to create electricity; and three, use some of this electricity to ignite a fossil fuel squirted inside the cylinder of an internal combustion engine. George Baldwin Selden redesigned George Brayton's two-cylinder gasoline engine in 1877 by creating a six-cylinder engine with an enclosed crank case. His engine weighed 370 pounds and produced three and half horsepower. The engine used crude naptha, a distillate (waste product) of the refining process.

Henry Ford got his first internal combustion engine running in 1893, but he didn't get around to putting the engine into an automobile until 1896. He sold the automobile and started making more. Ransom Olds started building automobiles in 1897.

Oilmen were relieved when people started buying the machines. They now had a market for the waste product that they'd been dumping down creeks and rivers for about forty years. They called it crude naptha. We call it gasoline.

AUTOMOBILES: NEW-FANGLED CONVEYANCE

Gasoline-powered automobiles started appearing on the streets in the 1890s. The noisy conveyance caught on; more than eight hundred automobiles were registered in the United States by the end of the decade. Eric L. Wilson of Bolivar owned the first automobile to appear on the Allegany County Oil Field. He purchased a new Stanley Stanhope Steamer, the one-seater, for $700 in 1900.

The Locomobile had wire wheels and pneumatic tires. The machine weighed four hundred pounds. Steam was supplied by a four-gallon copper boiler heated by a gasoline flame fed by a three-gallon tank. The dealer claimed that the car could go forty miles per hour and climb a 38 percent grade.

A crowd of curious onlookers gathered at the Bolivar railroad station on June 4, 1900, to see Wilson's automobile unloaded and prepared for an initial run. Several friends declined his offer to climb up and take a ride. However, after Wilson started, stopped and steered his car around a few street corners, Charles M. Van Curen, a local oilman, climbed up beside him and took a spin around town.

Wilson's driving ability amazed the onlookers. They didn't know that he had taken driving lessons in Rochester before buying the car. Soon, other people were asking for rides in his horseless carriage. On June 14, 1900, the *Bolivar Breeze* reported, "Eric L. Wilson and William L. Nichols rode to Olean last Friday in the new Locomobile, covering the eighteen miles in just an hour and a half."

Getting gasoline in Bolivar turned out to be a big problem, even though there were producing oil fields close at hand. Small amounts were available at the local drugstore, but Mr. Wilson solved his problem by ordering three barrels of gasoline from Olean's Acme refinery. Automobiles were a novelty. Nobody knew why or how they'd use an automobile or why they needed to buy one.

They were expensive, and most people lived within walking distance of where they worked. There were local stores and shops in every town. Local merchants could provide you with whatever you wanted to buy. Medical services, if available, were local. Railroads and trolleys could take you just about anywhere you wanted to go. Life was different, circumscribed by situation, tradition and location. When Henry Ford built his Model T, he had to find a reason for someone to buy it. About the only thing he came up with was, "It could take you to the country on a Sunday afternoon." Henry had a product, but he needed a reason for people to buy it.

When people started buying cars, it changed small towns. Cars changed cities, and now they've changed much of the world. Local gentry started buying cars, and then others did. Main streets, in most communities, went from being a gathering place for people, horses and wagons to being parking lots for automobiles. Trolleys were discarded; tracks were ripped up to make room for cars. Brick streets were covered with asphalt for a smoother ride. Filling stations, auto dealers, oil depots and the rest of the automobile infrastructure grew and expanded to displace existing services and technologies.

Cars have become a central feature of contemporary life; they exemplify mobility, status and freedom. Automobiles, aircraft and an expanding population generate increasing demands for energy. Technology drives demand. However, there have been unforeseen outcomes and consequences. Concern is being expressed about significant issues, some of which include the possibility of climate change, the safe recovery of fossil fuels and the availability of resources. Will there be adequate sources of energy to meet future needs?

The Millennium Project is a current independent nonprofit organization. It was founded following a three-year study by United Nations University, Smithsonian Institution and Futures Group International. The project recently predicted that global energy demand will triple over the next forty years. Meeting this demand will be a challenge.

NATURAL GAS, A FUTURE RESOURCE

Natural gas is being viewed as a partial solution. William Hart's original product is now available in nearby Sheridan, about seven miles from Fredonia, New York. Don Cotton owns several hundred gas wells in Chautauqua County and has opened a natural gas filling station, one of the few currently operating in the county. He's manufacturing and selling Compressed Natural Gas (CNG).

CNG is made by drying and compressing natural gas to more than three thousand pounds per square inch. It is stored in spherical or cylindrical containers. CNG has several advantages over gasoline when used in motor vehicles. It burns cleaner and produces more power with less pollution. Maintenance costs are reduced, and there is less fuel loss through leakage and evaporation.

Don Cotton switched all of his company's vehicles to natural gas about four years ago. Buffalo, New York, has converted delivery vehicles to natural gas, and several commercial carriers are also doing so. There is an ever-increasing demand for the cleaner-burning fuel. As of March 2011, there were 112,000 natural gas vehicles and more than one thousand CNG fueling stations in the United States. This is expected to grow.

General Electric is building natural gas–powered locomotives, units that can run on either natural gas or diesel. The western freight railroad company BNSF is shifting over to natural gas. Natural gas is cheaper, and once the conversion is completed, BNSF trains will have to refuel less often than diesel trains. Shell has entered the market.

Cotton well drilling in Sheridan, New York, provides compressed natural gas (CNG) to consumers. CNG can be used in place of gasoline, diesel fuel and propane. CNG produces fewer undesirable gases than other fuels, burns cleaner and is safer in the event of a spill. Natural gas, being lighter than air, disperses quickly. *Photograph by Hoffman.*

Cotton well drilling dries and compresses natural gas on site in an attached facility on the grounds in Sheridan, New York. There were 14.8 million natural gas vehicles operating in the world in 2011. There will probably be more. *Photograph by Hoffman.*

Dow Chemical, steel giant Nucor and Alcoa Aluminum are shifting over to using natural gas. Manufacturing requires massive amounts of energy; cheap, abundant natural gas is providing the incentive for these companies to expand and continue producing their products in the United States.

Coal-fired power plants are shifting over to natural gas to lower production costs while complying with environmental standards. This is happening in Dunkirk, New York, where the coal-burning generating plant built in 1950 is being converted to natural gas. After the conversion is completed, the plant will generate 435 megawatts of power. Forty-nine other regional generating plants in the United States and several Canadian plants are also making this transition. Natural gas is cleaner, it's cheaper and it appears that there's a lot of it.

Much of this natural gas is coming from Marcellus shale, a stratified black rock underlying much of northeastern United States. It's one of the world's largest gas fields, and it's situated to serve one of the world's largest energy markets: New York City, Philadelphia, Pittsburgh, Buffalo, Cleveland and Boston.

NRG Energy owns the steam generating station in Dunkirk. It is being converted from a coal-burning facility to natural gas and increasing its energy output to 435 megawatts. *Photograph by Niles Dening.*

Marcellus shale was originally mud at the bottom of a vast inland sea between 300 million and 400 million years ago. This mud was loaded with sediment, much of it being decaying organic material falling from the sea and other material washed into the sea. This material was compressed over millions of years, becoming layers of rock. Outcroppings of this rock can be seen in Western New York; however, these outcropping are just a few visible tips of this compressed sea of silt, clay and carbon that underlies much of the region.

Deep, dark bands of Marcellus shale run from the southern tier of New York State, through the western portion of Pennsylvania into the eastern half of Ohio and through West Virginia. Some geologists believe that Marcellus shale contains 170 trillion cubic feet of natural gas; others claim that it contains more than 500 trillion cubic feet, enough natural gas to power the United States for several years. It's considered a Super Giant Gas Field.

Pennsylvania is drilling, using horizontal drilling and fracking to release this gas. Pennsylvania companies drilled 27 wells in 2007, 161 wells in 2008, 785 wells in 2009, 1,386 wells in 2010 and 2,014 wells in 2011. New York State has been reluctant, drilling just 2 conventional vertical wells in Allegany County and 393 conventional vertical wells in Chautauqua County during this period, but it has banned horizontal drilling in Marcellus shale. Regulations are pending.

Utica shale is another large gas reserve; it's found a few thousand feet under the Marcellus shale. It's thicker than Marcellus shale and underlies most of New York State and much of Pennsylvania, eastern Ohio and West Virginia. Utica shale has been estimated to contain something between 5 and 25 billion barrels of oil and somewhere between 4 and 15 trillion cubic feet of natural gas—the largest hydrocarbon find in history.

Marcellus and Utica shale contain energy resources that have been projected to meet the nation's needs for the foreseeable future. Improvements in drilling technique with the passage of appropriate laws and regulation will secure that future. Efforts are underway to bring this about.

The progression of energy and light started in Western New York. It began when William Hart dropped his wife's washtub over bubbles rising to the surface of Canadaway Creek. Dr. Francis Brewer continued the process by sending a sample of rock oil to his chemistry professor at Dartmouth. Charles Brush moved it along by demonstrating his dynamo at Chautauqua. And Tesla tamed Niagara Falls, creating the electrical grid. Western New York is positioned and possesses the resources to continue play a significant role in the progression of energy and light.

BIBLIOGRAPHY

Barris, Lois, and Norwood Barris. *The Fredonia Gas Light and Water Works Company.* Research Monograph. Fredonia, NY: Barker Library, 1988.

Bitz, Robert W. *Transportation in Central New York and the Baldwinsville Area 1600 to 1940.* Baldwinsville, NY: Ward Bitz Publishing, 2012.

Boorstin, Daniel. *The Americans: The National Experience.* New York: Random House, 1965.

Brooks, Charles. *Frontier Settlement and Market Revolution: The Holland Land Purchase.* Ithaca, NY: Cornell University Press, 1996.

Buttrick, Tilly, Jr. *Voyages, Travels and Discoveries of Tilly Buttrick Jr.* Cleveland, OH: Arthur H. Clark Company, 1904.

Carlin, James, and Liz Austin Carlin. *Some Descendants of Richard Austin of Charlestown, Massachusetts, 1638.* Louisville, KY: Gateway Press Inc., 1998.

Clark, Christopher. *The Roots of Rural Capitalism: Western Massachusetts, 1780–1860.* Ithaca, NY: Cornell University Press, 1990.

Dolson, Hildegarde. *The Great Oildorado.* New York: Random House, 1959.

Downs, John P., and Fenwick Hadley. *History of Chautauqua County and Its People.* Boston: American Historical Society Inc., 1921.

Eaton, S.J. *Petroleum: A History of the Oil Region of Venango County, Pennsylvania.* Philadelphia: J.P. Skelly & Company, 1866.

Edson, Obed. *Biographical and Portrait Cyclopedia of Chautauqua County, New York.* Philadelphia: John M. Gresham & Company, 1891.

Ellis, Franklin, and Eugene Nash. *History of Cattaraugus County.* New York. Philadelphia: J.B. Lippincott & Company, 1879.

Evans, Estwick. *A Pedestrious Tour of Four Thousand Miles*. Concord, NH: Joseph C. Spear, 1819.

Everts, Louis H. *History of Cattaraugus County, New York*. Philadelphia: Press of J.B. Lippincott & Company, 1879.

Giddens, Paul H. *Early Days of Oil, a History Pictorial*. Princeton, NJ: Princeton University Press, 1948.

————. *Pennsylvania Petroleum, 1750–1872*. Titusville: Pennsylvania Historical and Museum Commission, 1947.

Grant, Kerry S. *The Rainbow City: Celebrating Light, Color, and Architecture at the Pan-American Exposition*. Buffalo, NY: Canisius College Press, 2001.

Henry, J.T. *The Early and Later History of Petroleum*. Philadelphia: James B. Rogers Company, 1873.

Herrick, John. *Bolivar, New York: Pioneer Oil Town*. Los Angeles, CA: Ward Ritchie Press, 1952.

————. *Empire Oil*. New York: Dodd, Mead & Company, 1949.

Ilisevich, Robert. *Remembering Crawford County: Pennsylvania's Last Frontier*. Charleston, SC: The History Press, 2008.

Jones, Jill. *Empires of Light*. New York: Random House, 2003.

Marcus, Alan, and Howard Segal. *Technology in America: A Brief History*. Boston: Cengage Learning, 1989.

Merrill, Arch. *Gaslights and Gingerbread*. New York: American Book-Stratford Press, 1959.

————. *River Ramble*. Rochester, NY: Democrat and Chronicle, 1945.

————. *Southern Tier*. New York: American Book-Stratford Press, 1948.

Messer, Jim. *Growing Up in the Bradford Oil Fields*. Bloomington, IN: Xlibris Corporation, 2008.

Miller, Sean K. *Pennsylvania's Oil Heritage: Stories from the Headache Post*. Charleston, SC: The History Press, 2008.

Minard, John S. *History of Allegany County, New York*. Alfred, NY: W.A. Ferguson & Company, 1896.

Sanford, Laura. *The History of Erie County, Pennsylvania*. Philadelphia: J.B. Lippincott & Company, 1894.

Schafer, Jim, and Mike Sajna. *The Allegheny River: Watershed of the Nation*. University Park: Pennsylvania State University Press, 1992.

Schivelbusch, Wolfgang. *Disenchanted Night: The Industrialization of Light in the Nineteenth Century*. Berkeley: University of California Press, 1995.

Seaburg, Carl, and Stanley Paterson. *The Ice King: Frederic Tudor and His Circle*. Boston: Massachusetts Historical Society, 2003.

Seifer, Marc. *Nikola Tesla: The Man Who Harnessed Niagara Falls*. Kingston, RI: MetaScience Publications, 1991.

Simpson, Jeffrey. *Chautauqua: An American Utopia*. New York: Henry N. Abrams Inc., 1999.

Speight, James G. *The Chemistry and Technology of Petroleum*. New York: Marcel Dekker Inc., 1991.

Turner, Orasmus. *Pioneer History of the Holland Purchase of Western New York*. Buffalo, NY: Jewett, Thomas & Company, 1849.

Waples, David A. *The Natural Gas Industry and Appalachia*. Jefferson, NC: McFarland & Company, 1958.

Whitney, Gordon G. *From Coastal Wilderness to Fruited Plain: A History of Environmental Change in Temperate North America from 1500 to Present*. New York, 1994.

Wilber, Tom. *Under the Surface: Fracking, Fortunes, and the Fate of Marcellus Shale*. Ithaca, NY: Cornell University Press. 2012.

Wilson, James. *The Earth Shall Weep: A History of Native America*. New York: Atlantic Monthly Press, 1998.

Young, Andrew. *History of Chautauqua County*. Buffalo, NY: Printing House of Matthews and Warren, 1875.

INDEX

ABOUT THE AUTHOR AND PHOTOGRAPHER

Douglas Houck grew up on a farm near Forestville, New York. He lived and worked in various communities in Western New York before moving to Maine, Hawaii and Florida. Doug and his wife, Nancy, now live in Westfield, New York, part of the year and Punta Gorda, Florida, the remainder of the year. Doug currently teaches English and writing courses at Edison State College and Barry University in southwest Florida.

Robert Hoffman (left), principal photographer, and Douglas Houck, the author, seated on the front steps of Doug and Nancy's house in Westfield, New York. *Photograph by Nancy Miller-Houck.*

Doug graduated from the State University of New York (SUNY) at Fredonia. He holds master's degrees in English from the SUNY College at Buffalo and public policy analysis from SUNY Buffalo. He completed a doctorate in English from SUNY Buffalo. Dr. Houck has published nine books—some fiction, others nonfiction—and wrote public policy papers in Maine and Hawaii. He is a member of the Chautauqua County Historical Society, the Peace River Center for Writers and Florida Writers Association.

Doug and Bob Hoffman are old friends. They collaborated on several Orleans County Bicentennial publications in 1976 while both worked for the Albion School District.

ROBERT HOFFMAN started taking pictures as a yearbook photographer at LaSalle High School in Niagara Falls, New York. He graduated from the State University of New York (SUNY) at Geneseo and did his graduate work at SUNY Brockport. Bob worked as a press photographer and served as the bureau chief for the *Albion Advertiser* in Albion, New York, from 1973 until 1976. He did the photography for the 1976 Orleans County Bicentennial calendar and history. Bob's photo of the Orleans County Courthouse graced the cover of the *New York State Bar* magazine.

Bob spent thirty-five years teaching history and social studies at the Albion High School in Albion, New York, before retiring in 2006. He has traveled the Great Lakes and Atlantic coast taking pictures of lighthouses in daylight and at nighttime. Bob has also participated in several photo charter trips, taking pictures of antique steam trains in New Mexico, Pennsylvania and New York.

Bob and his wife, Barbara, live in Medina, New York, with their two sons.

Visit us at
www.historypress.net

....

This title is also available as an e-book